Functional Applications of Text Analytics Systems

RIVER PUBLISHERS SERIES IN DOCUMENT ENGINEERING

Series Editors:

Steven Simske
Colorado State University, USA

Indexing: All books publishcd in this scries arc submitted to the Web of Science Book Citation Index (BkCI), to SCOPUS, to CrossRef and to Google Scholar for evaluation and indexing.

Document engineering is an interdisciplinary set of processes and systems concerned with the analysis, design, development, evaluation, implementation, management, and/or use of documents and document corpora in order to improve their value to their users. In the era of the Internet, the millennia-old concept of a document is rapidly evolving due to the ease of document aggregation, editing, re-purposing, and reuse. In this series of books, we aim to provide the reader with a comprehensive understanding of the tools, technologies, and talents required to engineer modern documents.

Individual documents include web pages and the traditional print-based or print-inspired pages found in books, magazines, and pamphlets. Document corpora include sets of these documents, in addition to novel combinations and re-combinations of document elements such as mash-ups, linked sets of documents, summarizations, and search results. In our first set of books on document engineering, we will cover document structure, formatting, and layout; document structure; summarization; and classification. This set will provide the reader with the basis from which to go forward to more advanced applications of documents and corpora in subsequent years of the series.

The books are intended to cover a wide gamut of document engineering practices and principles, and as such will be suitable for senior undergraduate students when using the first 2/3 of each book (core topics), and be extensible to graduate students, researchers and professionals when the latter 1/3 of each book is also considered (advanced topics). Students and graduates of data analytics, information science, library science, data mining, and knowledge discovery will benefit from the book series.

Functional Applications of Text Analytics Systems

Editors

Steven Simske

Colorado State University, USA

Marie Vans

Research Scientist, HP Inc. USA

LONDON AND NEW YORK

Published 2021 by River Publishers
River Publishers
Alsbjergvej 10, 9260 Gistrup, Denmark
www.riverpublishers.com

Distributed exclusively by Routledge

4 Park Square, Milton Park, Abingdon, Oxon OX14 4RN
605 Third Avenue, New York, NY 10158

First published in paperback 2024

Functional Applications of Text Analytics Systems / by Steven Simske, Marie Vans.

Routledge is an imprint of the Taylor & Francis Group, an informa business

Publisher's Note
The publisher has gone to great lengths to ensure the quality of this reprint but points out that some imperfections in the original copies may be apparent.

While every effort is made to provide dependable information, the publisher, authors, and editors cannot be held responsible for any errors or omissions.

ISBN: 978-87-7022-343-0 (hbk)
ISBN: 978-87-7004-319-9 (pbk)
ISBN: 978-1-003-33822-2 (ebk)

DOI: 10.1201/9781003338222

Contents

Preface

Text analytics do not exist in a vacuum: there must be a reason to perform them. Without context for the analysis, we provide data extraction and tabulation rather than any meaningful interpretation of the data. Thus, optimization in text analytics must be contextual. The central thesis of this book is that the most meaningful measure of the value of a chosen process for text analytics is output of the analytics that is more useful to the users than any of the other outputs available to them. This functional approach acknowledges that text analytic systems need to be learning systems: organic, dynamic, and both resilient and responsive to the changing conditions in which they operate. This, in turn, argues against the use of purpose-collected, human ground-truthed training and validation data. First off, training is time-consuming and expensive. Second, even humans with significant domain expertise in a field quite often disagree with the specifics of what training-derived metadata to assign. Third, training data can act as a rigidizing agent for the future training of a system; that is, it creates "asset inertia" that a system owner may be reticent to relinquish. Training data may age quickly, no longer representing the actual input domain of the system and/or it may have been designed for a specification that has since proven unsuitable – or at least in need of alteration. In this book, we are concerned with identifying alternative manners in which to train text analytic systems. We develop systems with two or more simultaneous output goals, with the tradeoff between the functioning of the two systems used to hone the specifications and settings of them both. This use of functional applications for the training of text analytics systems is shown to be broadly applicable to fields as diverse as summarization, clustering, classification, and translation.

Acknowledgement

Steve Simske's Acknowledgement:
What a delight this stage of the book is! Getting to thank even a fraction of the people who supported me on this journey. Thanks, Marie, for being a perfect person and friend to co-author this book with. In addition to believing in the subject matter, you look at every project differently than I do, allowing the work to grow. Thank you Marie! I would like to thank my support network, which was especially important in this very, very strange year. Thanks to Nicki Dennis at River Publishers for her tireless energy and perpetual encouragement. Thanks to Douglas Heins, Ellis Gayles, Tom Schmeister, Dave Price, Dave Wright, Steve Siatczynski, Dave Barry, Bob Ulichney, Matt Gaubatz, Igor Boyko, Jeff Ewing, Marty Esquibel, Pantelis Katsiaris, Mike Beiter, and Lee and Michelle Lahti for surprising me with unexpected grace, mirth, or both. Thanks to John Keogh, Reed Ayers, Gary Dispoto, and many, many other friends who have always been there for me. Thanks to my Thursday morning brainiacs, Paul Ellingstad, Mick Keyes, Gary Moloney, and Margie Sherlock for keeping me informed. Thanks to Ethan Munson, Dick Bulterman, Alexandra Bonnici, Cerstin Mahlow, Steve Bagley, Dave Brailsford, Charles Nicholas, Frank Tompa, Tamir Hassan, Matt Hardy, Evangelos Milios, Angelo De Iorio, Sonja Schimmler, Badarinath Shantharam, and the rest of my ACM friends for supporting this effort and me! Thanks to my campus friends, Ángel Gonzalez-Ordiano, Doncey Albin, Sam Wolyn, Wes Anderson, Sean Lahti, Naomi Andre, Adam Morrone, Ann Batchelor, Bo Marzolf, Erika Miller, Chrissy Charny, Ingrid Bridge, Mary Gomez, Katharyn Peterman, Dan Herber, Kamran Shahroudi, Jim Cale, Steve Conrad, Jeremy Daily, Tom Bradley, Jason Quinn, Brad Reisfeld, Chris Kawcak, John Macdonald, Leo Vijayasarathy, Valorie LeFebre, Jordan Foster, Ellen Brennan-Pierce, Luke Loetscher, Todd Gaines, and many others who have always been there to lend a hand. Thanks to Jon Kellar, Grant Crawford, Sandy Dunn, Suzanne Grinnan, Robin Jenkin, Scott Silence, Werner Zapka, and the rest of my IS&T friends. Thanks to my numerous online graduate students, who, combined, have the intellectual firepower to deplete Stockholm of its precious medals. Thanks to any and all I have regrettably forgotten to mention at the moment. Finally, thanks, most of all, to Tess, Kieran, and Dallen, who let me work on this with all my data

sprawled about the kitchen table with literally no end in sight. Thanks just for being who you are, functionally the very best of people!

Marie Vans' Acknowledgement:
Thank you Steve for the opportunity to work with you on this textbook and for helping me escape all those years ago so that I continue to work with you, learn from you, and, most importantly, laugh with you. Thank you too, for allowing me to sit in on your classes, supporting me through yet another degree and, in general, supporting my continuing education addiction. Thanks to Tess for the healthy snacks as well as the invaluable recommendations and advice and thanks to Kieran and Dallen for putting up with my weekly Friday afternoon invasion of their home. Thanks to my dear friend Deanna Krausse for all of her encouragement and for believing in me no matter what. Thanks to my husband, Fauzi, for bearing with the evening and weekend meetings and making sure the kids were where they needed to be when they needed to be there. Thanks too for extraordinary effort put into making sure we felt like we were eating out at a 5-star restaurant every single night. Thanks to my kids, Katy, Sumar, and Sami for being the greatest joy of my life and the justification for why I worked on this book.

List of Figures

List of Tables

List of Abbreviations

ABM	Agent Based Model
AHC	Agglomerative hierarchical clustering
ANN	Artificial neural network
ANOVA	Analysis of variance
AR	Augmented reality
ASCII	American standard code for information interchange
BLEU	Bilingual evaluation understudy
BTW	By the way
CAPTCHA	Completely automated public Turing test to tell computers and humans apart
CBRD	Currently-being-read document
CF	Collaborative filtering
CNN	Convolutional neural network
CONOPS	Concept of operations
COV	Coefficient of variation, or the standard deviation divided by the mean
CSV	Comma-separated value
DF	Degrees of freedom
DFB	Degrees of freedom between clusters
DFT	Degrees of freedom total
DFW	Degrees of freedom within clusters
EFIGS	English, French, Italian, German, Spanish
EM	Expectation−maximization
FIPS	French, Italian, Portuguese, Spanish
MS	Mean squared error
MSB	Mean squared error between clusters
MST	Mean squared error, total
MSW	Mean squared error within clusters
MT	Machine translation
NASA	National Aeronautics and Space Administration
ND	Next document
NER	Named entity recognition
NLP	Natural language processing

NMT	Neural machine translation
NN	Neural network
NNLM	Neural network language model
Norm_corr	Normalized cross-correlation
OCR	Optical character recognition
PMQ	Percentage of queries
POS	Parts of speech
RHW	Relative human weighting
RKMC	Regularized k-means clustering
RMS	Ranked matching score
RNG	Random Number Generator
RNN	Recurrent neural network
ROE	Reading order engine
ROI	Return on Investment
ROUGE	Recall-oriented understudy for gisting evaluation
SCUBA	Self-contained underwater breathing apparatus
SECOPS	Security + operations
SEM	Standard error of the mean
SS	Sum squared
SSB	Sum squared error between clusters
SSE	Sum squared error
SST	Sum squared error, total
SSW	Sum squared error within clusters
STD	Standard deviation
SVM	Support vector machine
TF*IDF	Term frequency times the inverse of the document frequency
UNL	Universal networking language
VR	Virtual reality
VSM	Vector space model
Word_corr	Normalized cross-correlation of two word sets

1

Linguistics and NLP

"To speak a language is to take on a world, a culture"
– Frantz Fanon

"The past is always tense, the future perfect"
– Zadie Smith

"Language disguises thought"
– Ludwig Wittgenstein

"I like you; your eyes are full of language"
– Anne Sexton

Abstract

In this chapter, the field of linguistics, or the study of language, is overviewed from the perspective of functional text analytics. Natural language processing (NLP) elements are discussed, and the means to compute important primary and secondary statistics about text elements are reviewed. Morphology, syntax, and semantics are reviewed briefly, and the Gestalt inherent in fluency is described. In the machine learning section, competition−cooperation is posed as a "twist" on expectation–maximization for optimizing the organization of text. Sensitivity analysis, agent-based modeling, and other means of functionally analyzing text are then covered, along with evolution-inspired approach of using language margins to drive test and measurement.

1.1 Introduction

Text analytics do not exist in a vacuum. There must be a reason to perform them. If there is no context for the analysis, we really only provide data extraction, resulting in tabulation rather than any meaningful interpretation of the data. Text analytics are performed on text in a wide manner of approaches; for example, text can be viewed as a bag of words, in which all the rich sequencing information is lost, or in its fluid form, in which case the final histograms of words are far less important than the development of ideas and, for example, the suggested reading order of the content. Thus, there is no single "best" way to analyze a library (meaning a large corpus of text-rich content). Optimization in text analytics must be contextual since content-based methods tend to emphasize specific rules or attributes at the expense of holistic understanding.

To counter the indecisiveness that often arises when choosing the specifics of the text analytics for a given project, we propose in this chapter, and in this book overall, the *functional* approach to designing, building, testing, measuring, validating, and improving text analytic approaches. Functional means that the settings are determined by finding out what works best in the context of one or more other tasks. If we wish to find out the best way to summarize some text, then we may wish to tie the summary to a learning program: a group of students who get the best score on a quiz after having used Summarizer A or Summarizer B (assuming there are proper controls between the two groups of students) provide a cogent argument for one summarizer versus the other. The people who are able to get the dinner they ordered in a French restaurant speak for value of Translator A versus the group of people choking down a meal they do not like after using Translator B. The stockbroker who is able to sift through reports and make the right investment decision because of the bulleted list of trends given to them by Text Analyzer A, when she cannot make heads or tails of the investment opportunities based on the output of Text Analyzer B, speaks for the use of Text Analyzer A. These results justify the design choices made **irrespective** of what types of analytical scores are given to the Summarizers, Translators, or Analyzers. In such a situation, in fact, we may wish to reinvestigate the analytical scores if they are not found to be in harmony with the functional utility of the outputs. **The central thesis of this book is that the most meaningful measure of the value of a chosen process for text analytics is that the output of the analytics are more useful to the users than any of the other outputs available to them**. Not that this should raise any warning flags: text analytic systems need to be learning systems. They need to be organic, dynamic, and both resilient and responsive to changing conditions in which they operate.

All of the requirements in the preceding sentence argue against the use of purpose-collected, human ground truthed training and validation data. First off, ground truthing is time-consuming and expensive. Second, even humans with significant domain expertise in a field quite often disagree with the ground truthing metadata to assign. Third, ground truthing can act as a rigidizing agent for future training of a system; that is, an expense that creates "asset inertia" that a system owner may be reticent to relinquish.

This is unfortunate since a fourth argument against ground truthed data is the fact that it may no longer represent the actual input domain of the system and/or it may have been designed for a specification that has since been found unsuitable or at least in need of alteration.

For the reasons cited with respect to ground truthing, in this book, we are concerned with identifying alternative manners in which to train text analytic systems. Ideally, we will simultaneously develop systems with two or more output goals and a tradeoff between the functioning of the two systems to hone the specifications and settings of both of them. At the simplest level, playing off two functionalities involves keeping one system's settings static while the settings of the second system are varied and noting which settings of the second system provide optimum output of the first system. Those settings are then held steady while the settings of the first system are varied, noting where the optimum output of the second system is achieved. This is, of course, an intentionally simple example: it is important to note that it largely assumes that the two systems are in the neighborhood of their global optimum output. However, it does illustrate the basic concept of a functional approach to defining a text analytics system.

Putting the simple system in the previous paragraph into a concrete example, suppose that we are using a linguistic system to simultaneously recognize different American English accents and, at the same time, to translate English expressions into Spanish for the more than 40 million United States citizens who speak Spanish as a first language. The set of iterations might look like the following.

(1) The settings for Dialect and Translation are each normalized to a range of $\{0.0, 1.0\}$, and each of 121 possible combinations of settings with a step distance of 0.1 between them is attempted; that is, $\{0.0, 0.0\}$, $(0.0, 0.1\}, \ldots, \{0.0, 1.0\}, (0.1, 0.0\}, \ldots, \{0.1, 1.0\}, \ldots, \{1.0, 1.0\}$. The results for Dialect and Translation recognition are computed for each of the 121 combinations.
(2) A Dialect setting of 0.7 and Translation setting of 0.5 are found to provide an overall accuracy of 65% for Dialect and 70% for Translation (mean = 67.5%). This was the best overall accuracy for the 121 combinations tested.
(3) The Dialect setting is then held at 0.7 and the Translation setting is varied from 0.4 to 0.6 with an incremental step size of 0.01. When the Translation setting is 0.46, the Dialect accuracy is 64% and the Translation accuracy is 77%. This is a new overall optimum, where (64% + 77%)/2 = 70.5%; so we stop there and move to step (4).
(4) The Translation setting is held at 0.46, and the Dialect setting is varied from 0.6 to 0.8 with an incremental step size of 0.01. At the Dialect

setting of 0.75, the Dialect accuracy is 71% and the Translation accuracy is 76%. This is a mean of 73.5%, a new optimum.
(5) We repeat next with Dialect setting held at 0.75 and Translation setting varied from 0.36 to 0.56, which yields (70%, 78%) at (0.75, 0.39). We continue these iterations until no further improvement, oscillatory behavior, or more than N iterations are observed. Our final setting is (0.74, 0.41) after three full iterations of holding Dialect and then Translation settings constant, which gives Dialect accuracy of 72% and Translation accuracy of 78%.

This example shows a significant reduction in error rate for both the Dialect (accuracy improved from 65% to 72%, and error rate was reduced by 20% of the original 35% error) and the Translation (accuracy improved from 70% to 78%, and error rate was reduced by 27% of the original 30% error). These are fairly large reductions, but, in fairness, it is possible because of the large neighborhood of "convexity" around this particular optimum. In problem spaces with many more local optima, a smaller variation in one setting while the other setting is held constant may be indicated. Regardless, this form of sensitivity analysis (let us call it the *variable−constant, constant−variable* iterative method) shows how a functional method of optimizing a system can be performed on any set of input data. While accuracy is still assessed here, it is simply at the level of understanding the words in dialect and translating properly, each of which is a relatively simple ground truthing process (as in, *please say these words*).

Other forms of text analytics, such as text clustering, can benefit from even simpler functional approaches. Here, we leave ground truthing altogether and simply consider alternating measurements of cohesiveness. As one example, for the first iteration, text elements are assigned to clusters based on with which documents they have the greatest similarity (Equation (1.1)). Similarity is defined by the percentages of all words in the document; for word i, this percentage is designated as $p(i)$ in Equation (1.1). The sum of dot products of all $p(i)$ for document A and document B is the Similarity (A,B).

$$\text{Similarity}\,(A, B) = \sum_{n=1}^{N_{\text{unique words}}} p\,(i)_A p\,(i)_B. \qquad (1.1)$$

After clustering due to similarity, the second half of the iterative clustering method is to unassign the K percentage of documents that are farthest from the centroids of these clusters. K is typically 10%−40% depending on the number of clusters that form in the first step. If many clusters form, 10% will suffice; if a single cluster forms, 40% is likely a better choice. Percentages in between these are readily set based on the distribution of the distances from the centroids of the remaining clusters. Highly skewed or bimodal distributions will tip this value toward 40%, while Gaussian distributions will tip this value toward 10%. Regardless, the centroids of the clusters will now be redefined and a second method (shared key terms, similarity of titles, shared authors, etc.) can be used to assign each unassigned document to a pruned cluster. At the end of the iteration, another assessment of cluster number and form is done, and clusters are pruned in advance of the next iteration of "similarity-prune-second method-prune."

More variations on these types of approaches are provided later in this chapter. However, this section makes it clear that alternatives to expensive human ground truthing are both important and, fortunately, possible. Frantz Fanon noted that to "speak a language is to take on a world, a culture," and, for our purposes, this means that we may be able to use cultural aspects of different languages to provide appropriate functional means of performing text analytics in each language. The differences in the usage of articles in Japanese and English, for example, might change the way we use (or ignore) stop words in these two languages. The much larger set of cases in German in comparison to Spanish might lead us to use lemmatization differently for these two languages.

We consider the other three quotes to kick off this chapter in light of what has been discussed so far. Zadie Smith, in a play on verb tenses, notes that the "past is always tense, the future perfect," which can also be a good attitude for us to take on as we move forward in functional text analytics. Building systems based on functionality, we recognize the failures from previous iterations and use them to provide higher predictive power — a perfect future. The philosopher Wittgenstein notes that "language disguises thought," which is perhaps a further argument against traditional ground truthing. Since the thought behind the text is not perceptible, what is more important is the flow of conversation. Indeed, as Anne Sexton notes, the more important part of linguistics is to be full of language. Language fullness can be best assessed functionally.

1.2 General Considerations

One of the first considerations to make when entering a "specialized" data field, such as text analytics, is to ascertain how specialized the field is in comparison to each of the other specialized fields of data analytics. After all, digital analytics proceed in a manner similar to text analytics inasmuch as they are concerned with the analysis of binary strings; that is, sequences such as 10010001, 0011100110011, and 01010100111. The word "Lambent" in ASCII (American Standard Code for Information Interchange) is the sequence "76, 97, 109, 98, 101, 110, 116," or in binary form:

Table 1.1   Computation of entropy for a large set of English words for all 26 English characters (ignoring capitalization) and for the first letters of a large set of English words. Data in the table collected from several references [Corn04][Nov12][Ohlm58].

Letter	p(i)	$\log_2[p(i)]$	$-p(i) \times$ $\log_2[p(i)]$	p(i)	$\log_2[p(i)]$	$-p(i) \times$ $\log_2[p(i)]$
		All letters			**First letters**	
a	0.08167	−3.61405	0.29516	0.11682	−3.09764	0.36187
b	0.01492	−6.06661	0.09051	0.04434	−4.49525	0.19932
c	0.02782	−5.16773	0.14377	0.05238	−4.25484	0.22287
d	0.04253	−4.55538	0.19374	0.03174	−4.97755	0.15799
e	0.12702	−2.97687	0.37812	0.02799	−5.15894	0.14440
f	0.02228	−5.48811	0.12228	0.04027	−4.63415	0.18662
g	0.02015	−5.63308	0.11351	0.01642	−5.9284	0.09734
h	0.06094	−4.03647	0.24598	0.042	−4.57347	0.19209
i	0.06966	−3.84353	0.26774	0.07294	−3.77715	0.27551
j	0.00153	−9.35225	0.01431	0.00511	−7.61246	0.03890
k	0.00772	−7.01718	0.05417	0.00856	−6.86817	0.05879
l	0.04025	−4.63487	0.18655	0.02415	−5.37183	0.12973
m	0.02406	−5.37722	0.12938	0.03826	−4.70802	0.18013
n	0.06749	−3.88918	0.26248	0.02284	−5.45229	0.12453
o	0.07507	−3.73562	0.28043	0.07631	−3.71198	0.28326
p	0.01929	−5.696	0.10988	0.04319	−4.53316	0.19579
q	0.00095	−10.0398	0.00954	0.00222	−8.81522	0.01957
r	0.05987	−4.06202	0.24319	0.02826	−5.14509	0.14540
s	0.06327	−3.98233	0.25196	0.06686	−3.90271	0.26094
t	0.09056	−3.46498	0.31379	0.15978	−2.64584	0.42275
u	0.02758	−5.18023	0.14287	0.01183	−6.40141	0.07573
v	0.00978	−6.67595	0.06529	0.00824	−6.92314	0.05705
w	0.0236	−5.40507	0.12756	0.05497	−4.18521	0.23006
x	0.0015	−9.38082	0.01407	0.00045	−11.1178	0.00500
y	0.01974	−5.66273	0.11178	0.00763	−7.0341	0.05367
z	0.00074	−10.4002	0.00770	0.00045	−11.1178	0.00500
SUM	**1.000**	**−145.338**	**4.176**	**1.004**	**−146.444**	**4.124**

"01001100,01100001,01101101,01100010,01100101,01101110,01110100."

Thus, a string of text can be directly converted into a string of binary digits (or "bits"). For language-based text – text like you are reading right now – there is less entropy in digits than that would be expected of a random signal. Almost two centuries ago, this unequal distribution of letters in a language was used in a pioneering detective story by Edgar Allan Poe [Poe43]. English language letter distributions have been tabulated for several decades, and we collect some of these results in Table 1.1 [Corn04][Nov12][Ohlm58].

Table 1.1 tabulates the frequency, $p(i)$, for the 26 English characters, along with the calculations necessary to determine the entropy of the language (Equation (1.2)).

$$\text{Entropy} = - \sum_{i=1}^{n_{\text{characters}}} p(i) * \log_2 p(i). \qquad (1.2)$$

The maximum entropy is a measure of the information content of the signal. For a 26-character alphabet, the maximum entropy is when each character occurs 1/26 of the time or $p(i) = 0.03846$. The closest letter to this is "*l*" in all English text and "*m*" as the first letter in English words, as shown in Table 1.1. Overall, this maximum entropy is $\log_2(26)$ or 4.700. This means that if every English character had equal probability of occurring, the English language could be represented by 4.7 bits of information. Adding in the capital letters, 10 letters, and punctuation, and textual content quickly rises to a maximum value above 6 bits. Of course, the characters in English are not randomly (that is, equally) distributed, with the entropy values in Table 1.1 summing to 4.176 (all letters) and 4.124 (first letters). These are the equivalent entropies of having 18.1 and 17.4 characters, respectively, evenly distributed in the language and as first letters of words in the language. Thus, English (as with all other languages) can be lossless compressed by an appreciable percentage.

Returning to our comparison of linguistics to other disciplines of analytics, the entropy is readily comparable to a wide variety of time series data; for example, bioelectrical time series like the electrocardiogram (ECG) [Rutt79], electromyogram (EMG) [Lee11], or electroencephalogram (EEG) [Igna10]. Depending on the sampling rate and electrode sensitivities, language may have more or less entropy than these physiological time series events.

Text information, in fact, can be directly represented as a time series. This is illustrated in Figure 1.1, in which we present a graph of "This is a sample sentence to show entropy" (dependent variable, *y*-axis) plotted against character position (independent variable, *x*-axis) into its numerical equivalent where $a = 1$, $b = 2$, $\ldots$, $z = 26$, and then plotting the character place against the character value (for simplicity, spaces between words and capitalization are ignored).

The point of this discussion and the information presented in Table 1.1 and Figure 1.1 is that text analytics can indeed benefit from other disciplines of analytics. Figure 1.1, for example, can be analyzed using a fast Fourier transform (FFT) to determine a frequency signature of the text, which may be used to identify the author, the dialect, and other relevant information about the source text.

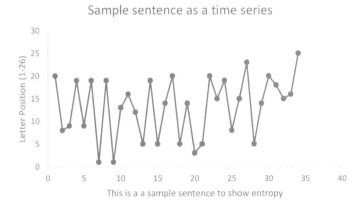

Figure 1.1   Graph illustrating the interconversion of text analytics into time series analytics. The curve comprises turning the sentence "This is a sample sentence to show entropy" into its numerical equivalent, where $a = 1$, $b = 2$, ..., $z = 26$, and then plotting the character place against the character value.

   With this introduction to text analysis, we turn to the more traditional means of analyzing text. To a linguist, important elements of text analytics include those that help uncover the structure of the language; that is, the morphology (or shape), syntax (rules and grammar), the phonetics (how do I speak individual words and liaison, elision, elide, and slide them together?), and the semantics (does "plethora" mean the same as "panoply"?). Linguistics is a massive field, in which text analytics (often labeled "computational linguistics") currently plays a small but important part. Currently, some of the broader components of linguistics – dialectology, historical-comparative linguistics, psycholinguistics, and sociolinguistics – are being recrafted in part in light of the Era of Big Data. From the perspective of this text, the field of linguistics, perhaps to be most influenced, is that of "applied linguistics" since our rather catholic definition of "functional text analytics" effectively extends to any and all applications of computational linguistics.
   Natural language processing (hereafter "NLP") is to linguistics what the calculator is to mathematics: it is the engine by which quantitative output can be generated to aid the linguist in her specific task, whether it is the use of linguistics for dialectology, psychology, sociology, or any other application. NLP, therefore, reaches from linguistics to shake hand with computer information systems, machine learning and artificial intelligence (AI), and human factors engineering to marry Big Data with linguistics.

Notable recent advancements in NLP have led to improved handwriting and other character recognition, speech recognition, conversational language understanding, and synthetic speech generation. Oh yeah, NLP has also helped to make the automatic call center a possibility. Sorry about that last one, but overall you have to admit that it opens up a lot of possibilities – not to mention leveling of the playing field – for people coming from a language, dialect, accent, or disability that leaves them outside of mainstream world languages (and IT support for those languages).

We now return to the topics of morphology, syntax, phonetics, and semantics as they relate to NLP. **Morphology** is the shape or form of things, and in linguistics, it focuses on the forms of words. The morphology of language includes the vocabulary, their tenses and cases, and their dependencies on surrounding words and other contextual cues to take their final forms. Prefixes and suffixes are integrally related to morphology, with the most "universal form" of a set of morphologically related words being called the stem. The stem is mapped to/from any of the prefixed and/or suffixed words with the same root word. For example, "maintain" is the stem for this set of 10 words: {maintain, maintainability, maintainable, maintained, maintainer, maintainers, maintaining, maintains, maintenance, maintenances}. However, it is not the stem for words related in meaning and similar in form (manipulate, manipulating, etc.) or related in meaning and origin (handle, handling, handlers, etc.). In Latin-derived languages, "mano" means "hand," and so handling and manipulating are essentially semantically identical. "Manipulating" means moving with a hand, and so it is also semantically highly synonymic with "handling" and "maintaining." We will discuss this more in a later section.

As seen from this example, singular and plural forms of a word share the same stem. However, a variety of ways in which plurals can be formed are also important to morphology. In English, most commonly, a plural form adds an "s" (cat → cats), but the vowels can change (mouse → mice), the consonants can change (calf → calves), and the plural may be the same as the singular form (moose). In English, and many other languages, at least some of these plural forms are irregular and must, therefore, be handled with an exception handler or look up table. This adds complexity to the task of the NLP analyst and programmer, but it is minor in comparison to the detection of case, tense, voice, or gender necessary in many NLP tasks. These aspects of NLP are termed "word inflection" and are focused on the assignment of different grammatical categories to each word in the text. In addition to those mentioned, inflection of verbs is termed "conjugation" and inflection of nouns, adjectives, adverbs, and, occasionally, other parts of speech (POS) (like the article "a" versus "an") are required.

Finally, spell checking is an important aspect of the morphology stage. Dictionaries of words can be used together with POS tables to determine the most likely word for each location, and Damerau−Levenshtein distance [Dame64][Leve66] is computed to determine the most likely candidates for typographical errors. This is also a good candidate for the application of machine learning processes. As with the rest of the morphology operators, spell checking is generally performed at the word level.

Syntax is the second large concern of NLP. Syntax addresses the parsing (POS determination) along with the rules and grammar of a language, which essentially aids in the construction of sentences as well as in their evaluation. Syntax can be tied to analytics directly through the POS determination: if the sequence of POS labels is non-sensical, it can be an indication of dialect, jargon, slang, or simply errors in the NLP analysis. Although, currently, syntax is analyzed at the sentence level, the advances in machine learning allow it to be pursued within a hierarchy, where information at one level (section; paragraph) is fed back to information at another (sentence; clause; phrase). Multiple candidates for each syntactical tagging can be attempted, based on the probabilities, until a best, or "consensus," output is reached. For example, if the statistics at the sentence level favor an adjective–verb tagging for two consecutive words with 84% confidence, but a noun–verb tagging with only 80% confidence, then we may wish to disambiguate them at the paragraph level. If the adjective–verb interpretation leads to 65% confidence in the paragraph, but the noun–verb interpretation leads to 68% confidence in the paragraph, we may decide that paragraph cohesion outweighs the sentence cohesion in this case. Most syntax analysis is performed at the sentence level, but it is likely to continue moving to the paragraph level as machine learning architectures continue becoming more sophisticated.

The third major NLP consideration is the **phonetics** of the language. These are the creation and interpretation of sounds in a language, at the letter (vowel and consonant), combinations (multiple vowels or consonants that provide their own sounds, like ch, sh, th, sp, ai, ea, oo, ay, etc.). The word "phonetic" in English is not phonetic since the first syllable is pronounced with the ubiquitous-to-English schwa sound. Phonetics are the rules, and exceptions to rules, of how the sounds in syllables and sequential syllable come together. Liaison and elision, the borrowing of a final sound in one word to lead the next and the omission of a sound for greater flow in speaking, are strongly incorporated into French, for example. In English, many elisions are directly indicated using contractions, such as in don't or it's (unpronounced "o" and "I," respectively). Phonemes are the smallest units investigated in NLP since they are the perceptually distinguishable sound "primitives" such as "b" and the schwa sound. Language phonetic rules build on the sequence of letters to provide pronunciation rules; for example, even though the "e" in "tape" is silent, this silent "e" creates a long "a" sound in "tape" which makes it completely different in pronunciation than "tap," with its short "a" sound. The phonetics of a language are obviously less important for text analytics in most circumstances; however, that is certainly not the case when dialect, jargon, slang, or plays on words are involved. A simple example of a play on words is an acronym, such as BTW (by the way). More complex acronyms, such as SCUBA (self-contained underwater breathing apparatus), require phonetic rules for people to be able to pronounce them; others, meanwhile, require at least some form of Gestalt to recognize the pronunciation, such as CAPTCHA (completely automated public Turing test to tell computers and humans apart).

The last, but by no means least, area of focus for NLP is the **semantics** of the text. Semantics are the meaning of the text expressions, which starts at the word level, and works upward to the phrase, the sentence, the paragraph, and the document level. Semantics, like other areas of NLP, are in a state of flux now as more advanced evolutionary, artificial neural network (ANN), and other machine learning capabilities are applied to NLP. Many areas of relevant semantics, though, benefit strongly from domain expertise input. For example, does "plethora" mean the same as "panoply"? From a semantics standpoint, the answer is probably yes, but from a functional standpoint, panoply may have more positive connotations, as it has a meaning of "impressive" in addition to its meaning as a multiplicity or collection. Plethora, on the other hand, is a large amount or collection of things, perhaps not impressive. These are subtle differences and, given that neither term is particularly common, may be very difficult for machine learning to disambiguate solely based on training data. The output of semantics is still evolving, but a type of classification called named entity recognition (NER) is a form of data mining involved in, effectively, tagging content with specific metadata. Also known by the names entity chunking, entity extraction, and entity identification, NER is a form of information extraction. Information extraction is the identification, or finding, of specific content, which, in this case, are named entities that appear somewhere in the (unstructured) text, that match pre-defined categories such as GPS locations, names, organizations, quantities, scientific units, currency types and amounts, and more. NERs benefit from a vocabulary of tokens, or tags, that can be applied, often organized as a taxonomy or ontology. These tokens often involve relatively sophisticated disambiguation of the possible categories for a term. One example is an article that discusses Google, Apple, and Microsoft. The NER annotations for Google and Microsoft are straightforward – that is, as [Organization] – however, Apple is more challenging since the taxonomy includes "apple" as a fruit and as part of the "Big Apple" (New York City). The assignment as [Organization] is decided instead of [object: fruit] based on statistical analysis of the context (not to mention the capitalization). The point being, semantics here produces metadata.

Another common output of semantics is relationships between named entities and other tagged information. The process of relationship extraction can be based on one or more of the following.

(1) Co-occurrence (two words or phrases occur near each other far more commonly than random chance would otherwise predict). We see iPad occurring with high probability in the same sentence or paragraph as Apple and, thus, deduce a relationship between them [Apple → iPad]. We can then use this established relationship to statistically "weight" the named entity relationship for Apple to that of [Organization] whenever the two terms are used with co-occurrence.

(2) Sequence (the two terms occur in order frequently — if directly one after the other, this can be used to define new compound words or catch phrases). Using the example above, we find the sequence "Apple iPad" with much higher probability than random sequential occurrence and, from this, further deduce the relationship [Apple → iPad]. We can then use this established sequential relationship, unsurprisingly, to further statistically "weight" the named entity relationship for Apple to that of [Organization] whenever the two terms are used in this sequence.

(3) Interchangeability (usually for nouns). Here, the two terms, if replaced by the same pronoun, seem to fit the same noun. This means the two terms have equivalent anaphoric resolution. As one example, consider the expression "Bob took his son, Bryan, out in the sailboat. Professor Ryan's little guy made him proud when he was able to tie a proper knot to keep the sail in place during the windstorm." To make this sentence free of anaphora, each pronoun is replaced by its proper noun: "Bob took his son, Bryan, out in the sailboat. Bob's little guy made Bob proud when Bryan was able to tie a proper knot to keep the sail in place during the windstorm." Here, we see that Professor Ryan is interchangeable with Bob.

Used efficaciously, then, semantic analysis leads to improved tagging, indexing, and categorization. Taking into account what semantics add, we return to the example of the stemming of the word "maintain": {maintain, maintainability, maintainable, maintained, maintainer, maintainers, maintaining, maintains, maintenance, maintenances} all map to the stem "maintain." Now, when semantics are brought into play, words that are related in meaning and spelling (manipulate, manipulating, etc.) or simply related in meaning and word origin (handle, handling, handlers, etc.) are deemed as semantically related to "maintain." Semantics are, therefore, truer to the linguistics and provide a deeper relationship between words than is possible for morphology, syntax, or phonetics.

Ultimately, linguistics and NLP come together at the level of fluency. This is the level of Gestalt (sudden, cross-topic insight to derive meaning) and pragmatics (being able to use the language functionally and effectively). For fluency, or linguistic Gestalt, we understand the text as a whole; that is, comprehensively. Within our minds, the running sum of recent conversational elements are simultaneously held in memory in a manner analogous to picturing a large set of geographical features simultaneously while navigating. Both the development of language [McCa19] and the development of map reading skills [Dill18] depend on the ability to "chunk"; that is, to aggregate information at a level intermediate to the primitive objects and the entire landscape (or language-scape). For maps, the primitives might be individual road sections, dots, rivers sections, and the like. For text, the primitives are the individual text characters and punctuation marks. The chunking for the maps are small connected geometries, e.g., triangles or rectangles connecting three or four intersections. The chunking for the text might be phrases, sentences, or even paragraphs. When you are learning a language, you may be translating every word, which can lead to the unfortunate misunderstandings that correspond with word-by-word translation. For example, imagine "wearing my heart on my sleeve" being translated into another language. Instead of being someone who shares their emotions readily, you are the perpetrator in a horror movie! Idiomatic expressions nicely identify the minimized size of chunking in language — without being able to chunk at this scale, we would really struggle to communicate. At the Gestalt level, the data analyst is concerned with generating output associated with communication. One type of this output is topic models. These are statistically generated models of the different subjects, or themes, in a document or collection of documents. Other advanced topics associated with Gestalt include document summarization, classification, and translation.

At this point, we have introduced some of the core topics of linguistics and NLP, which are the foundation blocks for text analytics. In each chapter of the book, as in this chapter, we provide an introductory and a general considerations section. After that, we consider machine learning aspects of the chapter's topics, continuing with design and system considerations. Then, applications and examples are considered. A test and configuration section completes each chapter. By providing these different views of each of seven key text analytics areas, we provide a broad functional view of the overall field of text analytics.

1.3 Machine Learning Aspects

1.3.1 Machine Learning Features

Machine learning approaches require features as input to generate rules, recognition, and other "content improvement" as output. When an element of text is considered as a "bag of words," for example, we lose much, if not all, of the sequencing information. However, simple feature generation from text can be an effective means of "triaging" text for later clustering, classification, categorization, and tagging, and so we cover these at least superficially here.

The first step in generating these generic, or "bag of words," features for text elements is to simply create a word count list. We prefer to use XML for this, but using a CSV (comma-separated value) file or a spreadsheet is acceptable. An example of word counts from a rather large document is shown here. Note that the word "the" is the most common with 5123 counts (occurrences), and the word "zzyzzx" (along with perhaps other words not shown) is the least common with 1 count.

```
<word_counts unique_words="5676" total_count="178053">
        <count name="the">5123</count>
        <count name="and">1356</count>
   . . .
        <count name="zzyzzx">1</count>
</word_counts>
```

As shown by the fields in the first line, unique_words="5676" total_count= "178053," we can see that for this particular document (a 600-page book), there are 5676 different words and a total of 178,053 words in the document. We start with the simplest pair of features to derive from the document as a "bag of words," which are the mean and standard deviation of the occurrences of the set of N_{words} words in the document. These are defined in Equation (1.3), where the mean is simply the total number of words – that is, the sum of occurrences of all the words occurring in the document, count(n) – divided by the total number of unique words, N_{words}. For our example, the mean = 178,053/5676 = 31.37. Thus, the mean number of times that one of the words actually in the document occurs in the document is just over 31 times. The standard deviation (σ) is then readily computed using the mean (μ) from Equation (1.3). This calculation is shown in Equation (1.4).

$$\text{Mean } (\mu) = \frac{\sum_{n=1}^{N_{words}} \text{count}(n)}{N_{words}} \tag{1.3}$$

$$\text{Standard Deviation } (\sigma) = \sqrt{\frac{\sum_{n=1}^{N_{words}} (\mu - \text{count}(n))^2}{N_{words} - 1}} \tag{1.4}$$

Note that the degrees of freedom, that is, the denominator term in Equation (1.4), are $N_{\text{words}} - 1$. For our example, the standard deviation was 25.64, meaning the range (31.37 − 25.64, 31.37 + 25.64) or (5.73, 57.01) is the range from $\mu - \sigma$ to $\mu + \sigma$. The ratio of the two values, σ/μ, is known as the coefficient of variance or COV. This value is computed in Equation (1.5).

Coefficient of Variance, COV (σ/μ)

$$= \sqrt{\frac{\sum_{n=1}^{N_{\text{words}}} (\mu - \text{count}(n))^2}{N_{\text{words}} - 1}} \Bigg/ \frac{\sum_{n=1}^{N_{\text{words}}} \text{count}(n)}{N_{\text{words}}}. \qquad (1.5)$$

As individual features, the triad $(\mu, \sigma,$ and COV) is not particularly useful for most text analytics. However, changes in σ are associated with changes in the effective vocabulary of the author. Changes in the mean, μ, are indicative of the length of the document, in general, and changes in COV tell us about the consistency of the word use throughout the document. A lower value of COV usually is indicative of a core vocabulary being used consistently, whereas a higher value of COV indicates different vocabulary sets being used at different times in the document. The latter, therefore, can be used as a test for multiple authors (or even plagiarism).

A less crude bag of words feature is based on the word histograms. The word histogram is simply the number of words that occur once, twice, three times, etc., in the document. This rearranges the <word_count> information from above in a manner shown here:

```
<word_histogram unique_words="5676" total_count="178053">
        <bin word_count="1">1143</bin>
        <bin word_count="2">317</bin>
    . . .
        <bin word_count="5123">1</bin>
</word_histogram>
```

Note that the word "zzyzzx" is one of 1143 words that occur only once in this document, while the word "the" is the only word that occurs 5123 times in the document. Altogether, we know that 178,053 words are in this corpus, and so the next thing to do is normalize the <word_histogram> information by dividing each term by 178,053. The results are shown here:

```
<normalized_word_histogram unique_words="5676" total_count="178053">
        <bin word_count="1">0.01765</bin>
        <bin word_count="2">0.00740</bin>
   ...
        <bin word_count="5123">0.0000056</bin>
</normalized_word_histogram>
```

There are far less bins for word count than there are bins for words since any two words that occur for the same number of times in the document write to the same bin count. It turns out that there are 396 different bins in this histogram. This normalized word histogram tells us a lot more about the distribution than the mean and standard deviation do. Here, we can see that the most common word, "the," occurs only 1.765% of the time. This is much higher than the expected value of 1/5676 or 0.018%. From this normalized word count, we can directly extract the probabilities, p_n, for each histogram bin n, where the bin corresponds to   <bin word_count=""> entry in the XML tabulation. In the above XML, p_1 = 0.01765, p_2=0.00740, ..., and p_{396}=0.0000056. The entropy of this "bins = word occurrences" histogram is given by Equation (1.6). The maximum entropy of such a histogram is $\log_2(n_{bins})$. In this particular example, the maximum entropy is 6.585 since $\log_2(396)$ = 6.585.

$$\text{Entropy}(\text{bins} = \text{word occurrences}) = -\sum_{n=1}^{n_{bins}} p_n * \log_2 p_n \qquad (1.6)$$

The actual entropy of the 396-bin data was 3.973, a substantial reduction in entropy. We will now show the computation of this entropy for a simple synthetic set of data to provide insight. Let us consider a 100-word document with 8 unique words, each of which occurs 12 or 13 times (4 words in each bin). The word count information is given by

```
<word_counts unique_words="8" total_count="100">
        <count name="word1">12</count>
        <count name="word2">12</count>
        <count name="word3">12</count>
        <count name="word4">12</count>
        <count name="word5">13</count>
        <count name="word6">13</count>
        <count name="word7">13</count>
        <count name="word8">13</count>
</word_counts>
```

The word histogram is then given by

```
<word_histogram unique_words="8" total_count="100">
        <bin word_count="4">100</bin>
</word_histogram>
```

In other words, word histogram bin for 4 has 2 entries, and all other entries in the word histogram are 0. This distribution results in the following normalized word histogram:

```
<normalized_word_histogram unique_words="8" total_count="100">
     <bin word_count="4">1.00</bin>
</normalized_word_histogram>
```

From Equation (1.6), this normalized word histogram distribution has entropy = 0.0.

Next, we consider a 100-word document with a different distribution, with 7 of the words occurring once and the other word occurring 93 times, as shown here:

```
<word_counts unique_words="8" total_count="100">
        <count name="word1">1</count>
        <count name="word2">1</count>
        <count name="word3">1</count>
        <count name="word4">1</count>
        <count name="word5">1</count>
        <count name="word6">1</count>
        <count name="word7">1</count>
        <count name="word8">93</count>
</word_counts>
```

This distribution results in the following normalized word histogram:

```
<word_histogram unique_words="8" total_count="100">
     <bin word_count="1">7</bin>
     <bin word_count="93">1</bin>
</word_histogram>
```

This distribution results in the following normalized word histogram:

```
<normalized_word_histogram unique_words="8" total_count="100">
     <bin word_count="1">0.07</bin>
     <bin word_count="93">0.93</bin>
</normalized_word_histogram>
```

The entropy of this normalized word histogram distribution is simply – $[(0.07) \times \log_2(0.07) + (0.93) \times \log_2(0.93)] = 0.366$. This is not a very large entropy, but it is non-zero.

In the third example, suppose we had the following word count distribution for a 100-word document with 8 unique words:

```
<word_counts unique_words="8" total_count="100">
    <count name="word1">5</count>
    <count name="word2">5</count>
    <count name="word3">10</count>
    <count name="word4">10</count>
    <count name="word5">10</count>
    <count name="word6">15</count>
    <count name="word7">15</count>
    <count name="word8">30</count>
</word_counts>
```

This distribution results in the following normalized word histogram:

```
<word_histogram unique_words="8" total_count="100">
    <bin word_count="1">1</bin>
    <bin word_count="2">4</bin>
    <bin word_count="3">3</bin>
</word_histogram>
```

This distribution results in the following normalized word histogram:

```
<normalized_word_histogram unique_words="8" total_count="100">
    <bin word_count="1">0.125</bin>
    <bin word_count="2">0.5</bin>
    <bin word_count="3">0.375</bin>
</normalized_word_histogram>
```

The entropy of this normalized word histogram distribution is simply – $[(0.125) \times \log_2(0.125) + (0.5) \times \log_2(0.5) + (0.375) \times \log_2(0.375)] = 1.406$. This is a much larger entropy than the previous two distributions. Designating this entropy e(nwhd) where *nwhd = normalized word histogram distribution*, we see e(nwhd) = 0.0, 0.366, and 1.406 for these examples.

It should be noted that another form of entropy can be computed for text distributions. This is the simple word frequency entropy, which we will designate e(wf) for comparison to e(nwhd). This entropy is much like the letter frequency entropy of Equation (1.2). This entropy is given in Equation (1.7).

$$\text{Entropy (word frequency)} = -\sum_{n=1}^{n_{\text{words}}} p_n * \log_2 p_n \qquad (1.7)$$

The probability p_n in this equation corresponds to the percentage of overall words that are word n. The maximum entropy is readily obtained as $\log_2$ (number of unique words) = 3.0 for these examples. We now compute this entropy for the three distributions above. For the first distribution:

```
<word_counts unique_words="8" total_count="100">
      <count name="word1">12</count>
      <count name="word2">12</count>
      <count name="word3">12</count>
      <count name="word4">12</count>
      <count name="word5">13</count>
      <count name="word6">13</count>
      <count name="word7">13</count>
      <count name="word8">13</count>
</word_counts>
```

The e(wf) for this distribution is unsurprisingly very close to 3.0 since the distribution is as uniform as possible for 8 bins and 100 words. Here, e(fw) = 2.99885. For the second distribution:

```
<word_counts unique_words="8" total_count="100">
      <count name="word1">1</count>
      <count name="word2">1</count>
      <count name="word3">1</count>
      <count name="word4">1</count>
      <count name="word5">1</count>
      <count name="word6">1</count>
      <count name="word7">1</count>
      <count name="word8">93</count>
</word_counts>
```

The value e(wf) = 0.56244, as low an entropy as possible when every word occurs at least once. For the final distribution:

```
<word_counts unique_words="8" total_count="100">
      <count name="word1">5</count>
      <count name="word2">5</count>
```

```
        <count name="word3">10</count>
        <count name="word4">10</count>
        <count name="word5">10</count>
        <count name="word6">15</count>
        <count name="word7">15</count>
        <count name="word8">30</count>
    </word_counts>
```

The value e(wf) = 2.77095, which is substantially less entropic than the first distribution and far less entropic than the second distribution. Note that these values for entropy do not correlate with the e(nwhd) values. These two entropies thus provide different insights into the word distributions.

Combined, the five features – μ, σ, COV, e(nwhd), and e(wf) – provide insight into the type of word "structure" we have in a text element, such as a document. However, this only scratches the surface of their value to the text analyst. In addition to computing these features on the whole document, the same set of five features can be computed on partitions of the document. These partitions can be sequential – e.g., the first $X\%$ of the document, where $X = 1/N$ of the document's length, create N partitions that can be used to generate up to $5(N + 1)(N/2) = 2.5(N^2 + N)$ features. This is because we have N partitions, $N - 1$ double-partitions (two consecutive partitions joined to formed one), $(N - 2)$ triple partitions, ..., and 1 N-length partition (which is the example given above since it was on the whole document or all N partitions). The number of total such individual or combined-sequential partitions is determined from Equation (1.8):

$$(N + 1)(\frac{N}{2}) = \sum_{n=1}^{N} (N - n + 1) \tag{1.8}$$

Multiplying this by the 5 features yields the $2.5(N^2 + N)$ features. Thus, a document with seven chapters, like this book, has 140 features from partitioning by chapter alone. Other forms of partitioning can be for POS (nouns, verbs, adjectives, and adverbs, mainly), by the lengths of the words, and other NLP factors. This partitioning can be performed on each of the sequential partitions described. Partitioning "logically" thus leads to another multiplication of the features useful as machine learning input.

As if this were not enough, normalized histograms can be computed for each of these partitions. This is performed by setting the total number of words to 1.0; in other words, by dividing the word counts for each word by the total number of words in the document. These normalized values allow more direct comparison of documents for the determination of text style, text authorship, and thus potential plagiarism. In addition, these normalized histograms can be subtracted or added for multiple documents to allow different aggregations of text to be compared. In this way, a document corpus can be pre-analyzed for similar authorship, allowing once anonymous documents to be assigned putative authorship once the identity of one document in the cluster's authors is established. This is definitely an exciting application of a rather prosaic set of features!

1.3.2 Other Machine Learning Approaches

Machine learning can be used to connect the two related fields of research of linguistics and NLP. They connect in the choice of algorithms for the primary NLP tasks (such as POS, tagging, categorization, and word sense) based on which language is identified. In other words, the language of the document is used as the means to select its analysis. The values of features, such as COV and entropy in the previous section, will generally vary from one language to the next. For example, articles are substantially different in comparing English (a, an, the) and Japanese (ga, ni, no, wa). A functional approach to linguistics and NLP is an iterative refinement algorithm. First, the language family can be identified from the character set. Next, the individual language is identified from the word counts. The dialect, if appropriate, can be identified from the rare terms within the language that occur with disproportionately high frequency in the document. This might include regional idiomatic expressions as well as variant spellings. Finally, jargon and slang dictionaries can be used to assign the document to specific specialties, trades, or other subcultures. All of these simple data extraction techniques lead to a wide feature set that can be used as training input for machine learning approaches as distinct as Bayesian methods and ANNs.

In this machine learning section, we are more concerned with the generation of useful features than we are with the already well-developed machine learning and AI algorithms and systems used to perform additional text understanding, such as translation, summarization, clustering, classification, and categorization (functional aspects of which will be covered in the following three chapters, of course). Why is this? Because algorithms and intelligent systems change over time. Meta-algorithmic [Sims13] approaches, among others, recognize this innate evanescence of the "state of art" by providing design patterns to build smarter systems agnostic to the details of the individual intelligence algorithms and systems. However, no intelligent system can perform its magic in a vacuum. The need for high quality input features is likely to never go away. Even neural network architectures acknowledge this through the addition of convolutional processes that effectively create millions – sometimes hundreds of millions – of "features" through their "tessellation" and recombination of input vectors.

This section, in effect, provides a process by which to convolve bag of words features into a much larger feature set, which can then be down-selected by the downstream intelligence system.

1.4 Design/System Considerations

1.4.1 Sensitivity Analysis

In each chapter of this book, the design/system considerations will be concerned with one or more adjacencies to the text analytic approaches that are the focus of the chapter. A key aspect of any system design is the sensitivity of the system to the settings employed. Sensitivity analysis is a set of procedures for determining if the design of a system is both stable (maintains its fidelity over time) and resilient (performs with minimal, if any, diminution of quality, accuracy, performance, and other measurements of interest in response to changes in the input). The latter is a function of the environment and, by definition, is speculative to imagine during the design phase. Thus, sensitivity analysis is (rightly) focused on the stability of the system, with the usually well-founded belief that stability is a good predictor for resiliency to the inevitable, and at least partly unpredictable, changes in the operating environment of the system.

For stability purposes, sensitivity analysis can be performed by holding a subset (usually all but one) of the settings constant while varying the rest of the settings (often just one) over a reasonable range. A simple example of that was given earlier in this chapter for optimizing both dialect understanding and translation accuracy in a text analytics system. This is an important starting point because, used properly, sensitivity analysis can be used as input for system optimization. This is not its goal, at least superficially, since we really are concerned with the stability of the system, and the optimized settings are necessary but not sufficient to ensure stabilization. That is, optimizing within a particular optimum (minimum or maximum) of the overall operation space does result in a local optimum, which is often the most stable point of operation within the local input space. However, it is not sufficient for overall system operation for at least two compelling reasons: (1) the local optimum space may not be smooth, symmetrical, or even reliable in all dimensions, and so the local optimum may be uncentered in this space such that a slight change in the input space dislodges the system from the local optimum and into another local space altogether; and (2) the local optimum space is not guaranteed to provide an optimum better than that in any other optimum spaces for the input. Each of these concerns, fortunately, can be addressed by approaches, such as Monte Carlo methods [Hamm64], which provide comprehensive or near-comprehensive coverage of the input space, allowing us to validate a suitable number of local optimum spaces to have relatively high confidence that one or more of these spaces lead to a global or near-global optimum.

It might sound like we are uncertain about this — is it not vacillating to keep saying "near-comprehensive," "high confidence," and "near-global" instead of just saying the best possible system? Actually, it would be irresponsible to say we have a global optimum for several reasons. First off, unless we perform an exhaustive search, in which every possible input is evaluated, and the best overall output is selected from this exhaustive search, we cannot be sure that the optimum was achieved. Second, the definition of "exhaustive" is built on quicksand for real-world problems. We take an innately analog world and digitize it. How many significant digits do we consider "exhaustive"? It turns out that it is pretty exhausting to even think about how to do an exhaustive search. Third, and perhaps most importantly, such an optimization over what generally amounts to an arbitrary input set (since we do not know how the input will change over time) is a potentially severe form of overfitting, meaning that whatever global optimum is found is unlikely to remain the global optimum in the face of new input. Combined, these reasons support the approach of looking for large spaces in which behavior leads to the same optimum. The larger the local neighborhood for optimization, the more likely that neighborhood is likely to remain a neighborhood for optimization when the data set is augmented.

1.4.2 Iterative Tradeoff in Approach

The iterative tradeoff approach is another variation on the expectation – maximization algorithm in which two different classes of text analytics are used in sequence, iteratively, until a consensus on the linguistics output is reached. Perhaps, these should be called two different phyla of algorithms since they might be used for classification, but the point is they are two (greatly) different approaches. The basic algorithm is as follows:

(1) perform analytic using approach 1 on the original 100% of the input;
(2) assign top 50% in confidence from approach 1 to their respective output (cluster, class, category, etc.);
(3) perform analytic using approach 2 on remaining 50% of the input;
(4) analyze the remaining 50% and assign the top 25% (half of the 50%) to their respective output (cluster, class, category, etc.);
(5) take the top 25% of the input most closely similar to the assigned 25% in (4) from either the 50% assigned in (2) or the 25% unassigned in (4) and assign with approach 2.

(6) whatever is not yet assigned (somewhere between 0% and 25% of the original input) now becomes the new 100% and restart the iteration in Step (1) above.

In the above, each match to output (cluster, class, category, or even most similar other element) comes with a "score" which is generally a confidence value. The confidences for all matches are ranked ordered (sorted) and the top 50% are readily distinguished from the remaining 50%. This is a good general purpose algorithm, and we have used it for clustering and classification. In clustering, we have used K-means and density-based spatial clustering of applications with noise (DBSCAN) [Schu17] as the two approaches with success. For classification, we have used the tradeoff of Bayesian, distribution based [Sims19], ANN, support vector machine (SVM), and other approaches for the two approaches. Note that the 50% and 25% are not required. Using a (33%, 33%, 33%) method, for example, also may provide good results, as may other settings, depending on the composition of the input (and the relative and absolute sizes of the different types of input). A particularly useful classification process involves using a boosting method [Freu99] as approach 1 and an ANN for approach 2.

1.4.3 Competition−Cooperation Algorithms

Conceptually related to the iterative tradeoff approach of the previous section are the competition−cooperation algorithms. Also iterative, this method uses a competitive algorithm in the first half of a cycle (one iteration comprises two half-cycles) and then uses a cooperative algorithm in the second half of the cycle. In this way, different levels of text aggregation are allowed to be measured prior to deciding the output. Importantly, this approach allows the comparison between the competitive step and the cooperative step to potentially argue for the creation of a new category. The basic algorithm is given here:

(1) Competition: find the single best match of a category to a sentence or other text element.
(2) Cooperation: find the best multi-category match to a sentence or other text element.
(3) Resolution: determine whether to tag the sentence (or other text element) with the single category, the multi-category, or create a new hyper-category comprising the single category + multi-category

The steps of this algorithm are relatively self-explanatory. In the first step, we compute the match for each text element — this could be matching it with a cluster, category, class, or another text element as in the previous section — and then determine a relevant confidence score. Next, we find the best multi-category match for the text element. This step gives us room for creativity in terms of how we define the multi-category: it can be the simple addition of two single categories or it can weigh the intersection between the individual categories differently than the non-intersecting content to create a different "centralizing" behavior. Regardless, we can then decide, based on the absolute and relative match of the text element with each of the three possible associations, {single category, multi-category, hyper-category} makes the most sense for the content. Suppose, for example, that the best match for a single category, S, is 0.75 and the best match for a multi-category, M, is 0.46. We label this $\{S = 0.75, M = 0.46\}$. We might then assign this element to S. Next, suppose that we obtain $\{S = 0.55, M = 0.54\}$. Since these two are close together, we may wish to assign it to M if M contains as one of its multi-categories the category S. However, if not, we are left to assign to S (highest value) or $H = S + M$, which is the new hyper-category containing single category S and multi-category M. If the match for H is at or above the value for S (e.g., match ≥ 0.55), we may assign to H, but if it is below, we may wish to assign to S and not create the new hyper-category. Once created in an iteration, we assign H as a new multi-category in the next iteration of the algorithm.

1.4.4 Top-Down and Bottom-Up Designs

Another interesting design approach is to mix together a top-down approach with a bottom-up approach. The top-down approach is, for example, the use of a taxonomy, an ontology, or a defined set of classes or categories, to which all input is assigned. A bottom-up approach, on the other hand, is usually statistically driven. A good example of a bottom-up approach is to use TF*IDF (term frequency times inverse document frequency) to identify outstanding terms in each text element. By outstanding, we mean terms that have a disproportionate frequency in those documents. We can then use those disproportionate terms as the features used for clustering. Cluster assignment via top-down approach is based on strict data mining and matching of terms in the documents against the terms assigned to the specific categories in the top-down definition. This top-down approach is generally relatively rigid in the features it uses for matching (a list of matches, and that is all). The bottom-up approach, however, can readily generate new features. This allows for both stability (consistent top-down behavior) and resiliency (the ability to add on new terms and features from the bottom-up statistics).

1.4.5 Agent-Based Models and Other Simulations

Simulations continue to gain momentum as tools for evaluating the manner in which networks of agents (or individual elements) work together effectively in a system. Agent-based modeling for linguistics [Stef95] was originally designed to help in large-scale linguistic tasks such as dictionary modification, grammars, and strategies of analysis. In agent-based modeling, the individual "patches" can be envisioned as sentences, paragraphs, or even documents, depending on the scale of the overall investigation. For example, if we are concerned with recommending a given reading order to a large set of documents, agent-based modeling can converge over time on recommended paths between these patches and, in so doing, provide novel learning plans, unique to the differential understanding of each reader, based on the decision-making heuristics, learning rules, and other adaptive processes internalized into the agent-based model (ABM). Since sequential reading, or reading order of material, can be guided by the content of what a reader already understands and/or has already read, each ABM simulation can be initialized with personalized settings and then be allowed to run to completion or confluence.

Additional simulations can also be performed using Monte Carlo methods, wherein the initial conditions are allowed to be randomly set over a reasonable (and quite possibly also personalized) range. Monte Carlo simulations can be run to generate conversational transitions that can then be judged by humans (or AI) for understandability. The rules generated from these simulations can be fed into AI algorithms as language features to assist in training.

1.5 Applications/Examples

The applications of linguistics and NLP are the subject of the rest of this book. The wide range of features that can be generated from text documents – from the mundane of mean, standard deviation, and entropy to the more sophisticated such as clustering, categorization, and classification – enable an even wider range of applications. One of the most important outputs of NLP is extractive summaries. Summarization is the topic of Chapter 2, and our focus is on extractive summarization since this is a more functional summarization approach than that of abstractive summarization. The former effectively selects down a text element to a kernel set of words, phrases, sentences, or other subelements that can function as a proxy for the often much longer original text. Abstractive summarization, on the other hand, tries to boil down a text element into a paraphrase of the original text that gets the gist across. Often, abstractive summarization changes the vocabulary, both absolutely (which words are used) and relatively (which words are used with higher frequency) in comparison to the original text. If you want to use somebody else's ideas in your (plagiarized) paper, you will want to use abstractive summarization since it will help your finished product deceive many of the common plagiarism-detection algorithms and software systems. The problems with abstractive summarization, however, are that these systems are often more difficult to understand, lose the point of the original text through the ambiguating process of abstraction, and cannot be indexed with as much veracity or at least accuracy since the defining terms of the original text may no longer be in the summary.

Such summarization troubles are unfortunate since other important applications for linguistics and NLP rely so heavily on the actual word occurrences of the original document. For example, generating a set of keywords for a document is an important task not only for tagging or indexing a document but also for associating a document with particular search queries to enrich the value of a corpus of text documents. For these tasks, extractive summaries usually provide better overall tagging performance. Moreover, as we will show in Chapter 2, functional approaches to extractive summarization can be used to ensure that the summaries are optimized to provide the best possible behavior on the set of search queries. Functional summarization, which compresses a set of documents in such a way as to make the behavior of the summaries on a set of search queries as close as possible as the behavior of the original (pre-summarized) documents, is an important example of explicitly functional means of performing text analytics.

Another important set of functional applications for text analytics are those of associating text elements with other text elements – clusters, classes, or categories – depending on the particular aspects of the problem. Summarization again makes a suitable vehicle for functional proofing of the machine intelligence algorithm. If the summarized versions of the document cluster together similarly; are classified with the same precision, recall, and confusion matrix behavior; or are assigned to the same relative categories, then we can *functionally declare* the summarization algorithm to be optimal. There is a lot to consider here. In some cases, we may decide that the set of summarized documents that provide the off-diagonal elements of the confusion matrix most similar to those of the non-summarized documents is optimal since the overall behavior of the mistakes in classification may be the most important aspects of a system. This is because confusion matrix-based patterns for machine intelligence, such as several meta-algorithmics [Sims13] and meta-analytics [Sims19] based approaches, may be the most resilient when the confusion (meaning class-to-class errors) is most similar.

Regardless, a variety of methods for functional assessment of text clustering, classification, and categorization will be provided in Chapter 3.

Translation is another key application of linguistics and NLP. Sometimes it is hard enough to communicate with another speaker of your mother tongue. Think about a fast-talking farmer from an Arkansas house, meeting a mumble-mouthed, Merseyside utterer of Scouse. Won't they need subtitles to know what each other is saying? Translation is a real-time issue, complicated not only by accents and dialects but also by specialized word usage such as slang, jargon, acronyms, plays on words, and *shudder* puns. Traditional means of assessing the accuracy of translation, therefore, may be less important than functional ones. In translation, what is most important is being understood, not in being translated word for word. Different means of ensuring translation accuracy through functional methods are described in Chapter 4.

The final application we will talk about here is text sequencing. This is extremely important in a learning, training, retraining, or continuing educational environment. Life is all about learning, and, thus, reading the right material precisely when you are optimally receptive to it is something worth pursuing. One specific area of research associated with text sequencing is reading order definition. Given everything I have read so far in my life, what is the best document for me to read next? Which one will simultaneously interest me, add to my knowledge and understanding, and also move me even closer to my specific learning or other reading-centered goals? This is not an easy topic to address, but, of course, it is of huge importance to anyone wishing to create a curriculum. A key aspect in determining a reading order is how to partition the influence of what you already have read from what you have not read. The degree of overlap of content is associated with what we call here the speed of learning. If we have a good estimate of how fast someone learns, than we can decide whether to use minimal overlap of "suggested next document to read" with previous material (fastest learners), intermediate overlap of "suggested next document to read" with previous material (normal learner), and maximum overlap of "suggested next document to read" with previous material (slowest learner). The latter classification is not insulting, as everyone will have different rates of learning for different types of content. Additionally, for continuing education, remedial learning, and for intensive studying such as it might accompany preparing for a certification examination, a normal or even fast learner may prefer a "slowest learner" setting for the purpose of really investing in the learning and/or memorization of the material. Much more will be said about learning in Chapter 6.

1.6 Test and Configuration

Linguistics and NLP, combined, cover a lot of ground. As we have seen, language is a huge area for machine learning and related intelligent system research. Fluency in a language is, for humans, sufficient to tell the fluency of other people. History is rife with examples of attempts at espionage and infiltration being thwarted because of insufficient pronunciation skills (primary linguistics skills) or the ability to calculate sums out loud (derived or adjacent linguistics skills). Although one might speak several languages, it is usual for one to perform language-adjacent skills, such as reciting the alphabet, counting, performing arithmetic, and giving directions, in one's primary language. Halting ability to perform these tasks in the secondary language is a known way to discover a spy or a mole. Language, therefore, is not simply for communication: it is also for organization, as skills as central to human existence as math, science, navigation, and memorization of lists are crucially dependent on one's linguistic abilities. As far as test and configuration is concerned, then, a person's performance in a given language on such adjacencies is an excellent test approach for true language fluency. It is not enough to be language-fluent; one must also be *language-adjacency fluent*. Sometimes this adjacent fluency is the difference between a successful reconnaissance mission and a firing squad; more prosaically, it is the difference between leaving an exorbitant and a simple generous tip in an overseas café.

In documents, language has a wide variety of roles. Language is associated with figure and table headings, references cited in the text, the flows of articles (e.g., continued on page X), and classification of different elements of text (title, section name, leading text, italics, underline, boldface, and the like). Such multi-sectioned text elements (and, in particular, one of our favorites being multi-article magazines) are an ideal set of text elements for testing and configuration optimization. Text analytic designers can measure, separately, both the success of primary linguistic applications (such as search ordering, clustering, classification, and document reading order) and secondary linguistic applications (such as header identification and reference understanding). This allows a linguistics system to be good at both data extraction (primary tasks) and knowledge generation (secondary tasks).

Another excellent opportunity for testing and configuring a linguistics and NLP system are the variety of specialized language usage vocabularies. These include abbreviations, acronyms, slang, jargon, and dialect. While there is some overlap among these sets of language, we describe them here in a way to allow each to stand on its own as a test set for configuring the overall system. Abbreviations are shortened forms of words, usually followed by a "period" punctuation in European languages. For example, "mgmt." is an abbreviation for "management" and "Ling." is an abbreviation for "Linguistics." Acronyms, on the other hand, are a subset of abbreviations in which the resulting compressed text (largely or completely) comprises the first letters of a meaningful sequence of text. As one example, "NASA" is an acronym for "National Aeronautics and Space Administration." If an abbreviation is a specialized form of language, then an acronym is a specialized form of language. Slang is another specialized form of language, which is usually informal language that is characteristic of a subset of the overall citizenry. Slang, if successful, is more generally adopted over time and becomes part of the common vernacular. Examples of now commonly adopted slang terms are "hangry" (so hungry you are angry) and "ain't" (isn't). Jargon, from our perspective, is more specialized than slang because it is more formal language; that is, it is a specialized vocabulary adopted by a subset of folks (e.g., a particular profession, organization, or group of hobbyists), and the vocabulary is generally esoteric to people not part of that particular circle. For example, "low-pass filtering," "differential amplifier," and "input impedance" might not mean much to you, but they certainly do to an electrical engineer. Jargon is not only specialized, it is essential. Imagine having to form compound words to explain input impedance: "MeasurementOfTheInverseOfTheRatioOfACurrentAppliedToTheTwoInput ElectrodesOfTheCircuitToTheVoltageMeasuredAtTheSameElectrodes." That is not jargon − that is garbage. Please, jargon away! The fifth form of specialized language in this section is dialect. A little bit different than the others, dialect does not need to be accompanied by a difference in spelling, in which case we can term this *latent dialect*. This is dialect that is written in identical fashion to any dialect. For example, a cowboy from Wyoming, a Texan from College Station, and a lumberjack from northern Minnesota might each say, "Howdy!" but they are likely to say this with three quite different accents. The set of all their differentiating accents of all of the words in the vocabulary is their dialect. Some linguists consider dialect the combination of accents and regionally varying vocabulary, but it is clear that a dialect transcends vocabulary no matter how specifically you define it.

All of these specialized language elements − abbreviation, acronym, slang, jargon, and dialect − have in common the fact that they drive change, eventually, in the overall language. Like evolution in general, wherein new species often differentiate in tidal pools cut off from the larger ocean or in pioneer communities that spend generations away from the larger gene pools, evolution in languages often comes from the margins. In machine learning, we can think of this "evolution on the margins" as a form of boosting in which the marginal linguistics contribute more to the overall language analytics than the more central terms. Our test for the significance of these specialized vocabularies is in reference to the more central terms that combined may be used to define the "normal" dialect of the language. The central terms compete with the specialized terms in defining what is normal and cooperate with them to define the future of the language. Thus, another form of competition−cooperation algorithm is naturally ongoing in the definition, and the redefinition, of what a language is.

1.7 Summary

This chapter highlights functional text analytics, starting from the crucial approach of simultaneously developing systems with two or more output goals, and trading off between the functioning of the two systems to hone the specifications and settings of them both. This was illustrated using dialect understanding trading off with translation accuracy. This sensitivity analysis approach was designated the *variable−constant, constant−variable* iterative method. Next, the ability to convert a text analytics problem into a digital (binary) or time series (e.g., physiological) analysis problem was described. We then explored the relationship between linguistics and NLP. Different aspects of NLP include morphology, syntax, phonetics, and semantics. All of these aspects combined equate to Gestalt, or fluency, in language. In the machine learning section, we demonstrated how a huge set of machine learning features can be derived from five simple calculations: mean, standard deviation, COV, and two forms of entropy. Important design and system considerations in linguistics and NLP include sensitivity analysis, in which system stability and resilience are simultaneously considered for system "optimization." Several iterative approaches related to sensitivity analysis – including iterative tradeoff of two approaches, the competition−cooperation algorithm and top-down, bottom-up designs – were then overviewed. Functional means of assessing summarization, clustering, classification, categorization, translation, and text reading order were then overviewed. Finally, we described the concept of language adjacency fluency, and some examples of specialized language usage (document elements, slang, jargon, and dialect) in the context of document test and measurement.

References

[Corn04]   Cornell University, "English Letter Frequency," available at http://pi.math.cornell.edu/~mec/2003-2004/cryptography/subs/frequencies.html, accessed on 5 July 2020 (2004).

[Dame64] Damerau FJ, "A technique for computer detection and correction of spelling errors," Communications of the ACM 7(3), pp. 171-176, 1964.

[Dill18] Dillon MR, Spelke ES, "From map reading to geometric intuitions," Developmental Psychology 54(7), pp. 1304-1316, 2018.

[Freu99] Freund Y, Schapire RE, "A short introduction to boosting," Journal of Japanese Society for Artificial Intelligence 14(5), 771-780, 1999.

[Hamm64] Hammersley J, "Monte Carlo Methods," Springer, Monographs on Statistics and Applied Probability, 178 pp., 1964.

[Igna10] Ignaccolo M, Latka M, Jernajczyk W, Grigolini P, West BJ, "The dynamics of EEG entropy," J Biol Phys 36(2), pp. 185-196, 2010.

[Lee11] Lee T-R, Kim YH, Sung PS, "A comparison of pain level and entropy changes following core stability exercise intervention," Med Sci Monit 17(7), pp. CR362-CR368, 2011.

[Leve66] Levenshtein VI, "Binary codes capable of correcting deletions, insertions, and reversals," Soviet Physics Doklady 10(8), pp. 707-710, 1966.

[McCa19] McCauley SM, Christiansen MH, "Language learning as language use: A cross-linguistic model of child language development," Psychological Review, 126(1), pp. 1–51, 2019.

[Norv12] Norvig P, "English Letter Frequency Counts: Mayzner Revisited or ETAOIN SRHLDCU," available at http://norvig.com/mayzner.html, accessed on 5 July 2020 (2012).

[Ohlm58] Ohlman HM, "Subject-Word Letter Frequencies with Applications to Superimposed Coding," Proceedings of the International Conference on Scientific Information, Washington, D.C., 1958.

[Poe43] Poe EA, "The Gold Bug," Philadelphia Dollar Newspaper, 1843.

[Rutt79] Ruttiman UE, Pipberger HV, "Compression of the ECG by Prediction of Interpolation," IEEE Transactions on Biomedical Engineering 26(11), pp. 613-623, 1979.

[Schu17] Schubert E, Sander J, Ester M, Kriegel HP, Xu X, "DBSCAB revisited, revisited: why and how you should (still) use DBSCAN,"ACM Transactions on Database Systems 42(3), 19:1-21, 2017.

[Sims13] Simske S, "Meta-Algorithmics: Patterns for Robust, Low-Cost, High-Quality Systems", Singapore, IEEE Press and Wiley, 2013.

[Sims19]   Simske S, "Meta-Analytics: Consensus Approaches and System Patterns for Data Analysis," Elsevier, Morgan Kaufmann, Burlington, MA, 2019.

[Stef95]   Stefanini M-H, Demazeau Y, "TALISMAN: A multi-agent system for natural language processing," SBIA 1995: Advances in Artificial Intelligence, pp. 312-322, 2005.

2

Summarization

"To summarize the summary of the summary: people are a problem"
– Douglas Adams, 1980

"Egotism is the source and summary of all faults and miseries"
– Thomas Carlyle

Abstract

Summarization is one of the more important functional tasks to accomplish in text analytics. From a broad perspective, summarization is effectively a variable repurposing approach for text. If we decide that the amount of compression is the most important factor in summarization, then, at its highest compression, summarization provides a set of keywords, labels, or indexing terms for a larger body of text. At the other extreme, the lowest compression level of summarization is an abridged text. In between are such familiar summaries as Spark Notes and CliffsNotes. In this chapter, we cover this wide range of summarization approaches and introduce the reader to both the statistical and the machine learning approaches of value in the functional compression of text; that is, summarization. Key to the chapter is the use of sentence weighting and co-occurrence approaches, along with functional means of assessing the relative goodness of the distinct summarization approaches.

2.1 Introduction

At its broadest level, there are two branches of summarization: abstractive and extractive. Of the two, extractive summarization is easier to describe since it consists of the selection of a subset of the information in the original document. This partial selection constitutes a compression of the content, but the compression is **lossless** in the sense that none of the text selected has been

35

altered from its original form. Abstractive summarization, on the other hand, involves a form of **lossy compression** inasmuch as the subset of the text that is selected comprises novel phrases, potentially using different words, word order, or phrases sequencing to provide a plainspoken summary along the lines of, ideally, conversational language. The abstractive summary has the advantage of, at least in theory, flowing better than an extractive summary. This advantage of abstractive summarization, however, comes with a number of cautions. The summary may not represent the style, tone, or voice of the original author. The summary may also introduce a different vocabulary than the author, with potentially negative effects on the later ability to perform text analytics such as indexing, clustering, categorization, and classification on the document.

One of the key issues to keep in mind about summarization is the quote by Douglas Adams; that is, that people are part of the solution and, thus, part of the problem. A machine learning algorithm might come up with a mathematically optimized solution that nevertheless reads poorly. For example, suppose the machine learning algorithm is based on the relative frequencies of words in a phrase compared to the broader language. The following sentence may be familiar to you:

"To be or not to be, that is the question." (Hamlet)

This sentence, one of the more famous sentences in all of literature, starts one of Hamlet's soliloquys and is, without doubt, one of the top five lines from the play. Yet, when we consider the 10 words in the phrase, they are all very common in the English language. The first 9 words are all in the top 50 words in the English language, and the last word, "question," is in the top 500. The mean score of the 10 words, from one online list, is 46. That is, the mean score for a word in the sentence is 46th in prevalence in the English language. There is little hope that, without additional context, this line will be chosen over almost any other line in the play. These words are common to the English language in general and not to the play in particular. Any line in Hamlet that includes one or more of the words "Hamlet," "Gertrude," "Ophelia," "Claudius," or for that matter "decide," "gold," or "possible" (these latter three rank above 460th in prevalence in English), is a phrase that will rank as more of an outlier phrase than the more famous line above. Thus, context, and not just content, drives what text should survive the functional compression that is summarization.

Carlyle's quote further elucidates the messy idiosyncrasy that usually accompanies summarization. Returning to Hamlet, the line "To be or not to be, that is the question" may be more salient to a philosophy major than the line:

"Neither a borrower nor a lender be." (Polonius)

However, this line may be more important to a historical linguist, who is capturing common phrases of Elizabethan England and who finds this and another phrase by Polonius to be more important:

"This above all: to thine own self be true." (Polonius)

Neither of these sentences stands out for the rareness of their terms ("thine" does a little more in 2020 than in 1600), meaning that like the earlier line spoken by Hamlet, these phrases are not particularly likely to stand out in a general machine learning algorithm's extractive summarization. Yet, to many modern readers, all three lines are instantly recognizable and usually listed among the top 10 phrases from Hamlet, which itself is usually listed among the top 10 pieces of literature from all of history. This is, therefore, a rather important contradiction to resolve: these three lines simply must be part of an extractive summarization of Hamlet. So, why are they not selected?

They are not selected because looking for their importance at the word level is looking for relevance incorrectly in both **scale** and in **structure**. The **scale** is wrong because these phrases, ranging from 7 to 10 words in length, are roughly an order of magnitude larger in scale than individual words, and so, when dissembled, they lose their integrity as phrases. Compound nouns, such as "New York City" and "cat of nine tails," are intermediate examples of the concept of scale. The **structure** is wrong because instead of looking for the relative rareness of the dissembled terms to determine which phrases allow an element of text to stand out from its peers, we are looking for the relative frequency of an expression across a wide diversity of other documents to weight it appropriately. The expression "to thine own self be true," for example, might be found on a yoga gym's web page, on a psychologist's Facebook page, or on someone's LinkedIn statement. Its ubiquity means it is not necessarily a differentiator for summarization today but probably should be in its original source. We, thus, see how very contextual − in **scale**, **structure**, and now **time** − summarization can be.

Given this high-level introduction to the breadth of summarization concerns, we next turn to an overview of some of the existing types of summarization. We start with a broad historical overview and then move to the work performed by the authors and colleagues over the past decade. We then address summarization in the context of overall text analytics.

2.2 General Considerations

2.2.1 Summarization Approaches – An Overview

In 1958, Hans Luhn saw that by using machine methods to automatically create abstracts for technical literature, a great deal of intellectual effort could be saved and be less influenced by the biases, backgrounds, opinions, and attitudes of the humans who normally produced abstracts for cataloguing purposes [Luhn58]. More than 50 years later, many automatic summarization techniques are still based on word-frequency and context (co-location, domain, corpus of related documents, etc.). The idea of assigning measure (weights) of significance to each of the sentences is still employed today. Other early works include those by Edmundson and Wyllys [Edm61, Edm69], Rush *et al.* [Rush71], Sparck-Jones [Jones93], and Kupiec [Kup95], the last two of which showed enough promise to set off new fascination for this field that makes it an active area of interest till today.

As we will be describing in greater detail both extractive and abstractive summarization, here, we will quickly define and show some examples by describing the work of other researchers using these two approaches.

Extractive Approaches:

Extractive approaches use text *as is* rather than generating new text for summarization. The very first extractive approach was used by Luhn [Luhn58], who defined a word-frequency-based measure using the idea that the more often an *important word* (*stop words* such as "the," "and," "or," etc. are not considered important in this context) appears in a sentence, the more significant the sentence must be and, therefore, should be included in the summary. In general, raw frequencies of words are not helpful, however, because as the document grows longer, the frequency of words will influence the significance measure [Basi16]. Word frequency can be normalized by document length, which can then be used as a weighting factor for significance. TF-IDF, or term frequency-inverse document frequency, is a related approach that we will be covering in more depth in the sections below.

Instead of looking at the frequencies of words to determine importance, there are several feature-based approaches that assume that certain characteristics of text identify significant sentences or phrases. For example, titles, abstracts, sentence position, proper nouns, and the like are indicators of relative importance [Basi16]. Such approaches may require additional methods such as document segmentation in order to identify these features.

Finally, supervised machine learning extractive techniques include Naïve Bayes classifiers, neural networks, and support vector machines (SVM) [Wong08]. Bayes classifiers determine the probability of a sentence being included in the summary based on a set of sentence features and the highest ranked sentences are chosen for the summary [Basi16]. In one neural network approach, summary sentences were used to identify features that best characterize a summary sentence [Kai04]. In another, a training set consisting of Cable Network News (CNN) articles was used to train a neural network to rank sentences, and the summaries consisted of the highest ranked sentences [Svor07].

Abstractive Approaches:

Abstractive approaches do not necessarily use the text as is but may paraphrase parts of the text or transform the text to build a summary. These approaches can suffer from several issues, including the production of inaccurate details and the repetition of information.

See *et al.* address both issues by using what they call a "pointer−generator network" [See17], which is built on top of a "sequence-to-sequence" model. This is a common model used for abstractive summarization that employs recurrent neural networks (RNNs). The pointer−generator network copies words from the source text (to address the inaccurate details) by pointing, and unseen words are produced using the generator. They also track the summarization as it is being produced in order to avoid repetition.

Nallapati *et al.* use an attentional encoder−decoder RNN for abstractive text summarization on two different IBM corpora [Nall16]. Using the decoder to "point" (copy) a word or encoder to "generate" (produce) a word addresses the issue that abstractive approaches encounter with respect to inaccurate details. Because the summary outputs from the encoder−decoder system tended to contain repetitive phrases, the researchers added a temporal attention model that essentially keeps track of which parts of the document it has already processed and discourages it from looking at those pieces again. We will see similarities of this approach in our regularization approaches later in the chapter.

Finally, Rush *et al.* used a different approach, called attention-based summarization, which joins a neural language model with an attention-based encoder. The language model consists of a feed-forward neural network language model (NNLM) used for estimating the contextual probability of the next word, while the encoder acts as a conditional summarization model [Rush15]. In general, each word of a summary is generated based on the input sentence. The goal of the experiments is to use the summarization model for generating headlines. Training occurs by pairing a headline with the first sentence of an article (rather than the entire article) to create an input-summary pair. Interestingly, by using *extractive tuning*, the authors are able to address the issue of inaccurate details (similar to decoding or pointing in the works cited above) by tuning with a small number of parameters after training the model.

2.2.2 Weighting Factors in Extractive Summarization

One advantage of extractive summarization over abstractive summarization lies in its ability to incorporate many different weighting schemes for phrases (usually sentences) that are selected for the final summary. In 2013, led by our Brazilian colleagues (especially the indefatigable Rafael Lins), we started researching the role of sentence scoring techniques in the accuracy of extractive text summarization [Ferr13]. A set of 15 different approaches were attempted. Here we define the basis for each approach and discuss some of the ways in which they can be used in future. Then we describe how they might be used together.

(1) **Word scoring** is the first method. Conceptually, this is the easiest possible means of ranking phrases and sentences. However, there are some nuances, and these provide a suite of possible approaches, making this method more useful than might be at first estimated. At the simplest level, word scoring maps a query expression to each potential matching phrase. Suppose that the query is "Pittsburgh, Pennsylvania" and we perform the simplest possible word scoring wherein the score is simply the sum of the terms "Pittsburgh" and "Pennsylvania." The following two sentences, therefore, have scores of 2.0 and 3.0, respectively:

> Pittsburgh, Pennsylvania, is where the Ohio River starts (Weight 1.0 + 1.0)

> Pittsburgh, Pennsylvania, is home of the Pittsburgh Steelers (Weight 2.0 + 1.0)

However, the relevance of a phrase should, most likely, increase non-linearly with the number of times each term matches the query. If a simple power law is used, then a matching term is weighted by the square of the number of matches. In this case, the two sentences above are weighted 2.0 and 5.0, respectively:

> Pittsburgh, Pennsylvania, is where the Ohio River starts (Weight 1.0 + 1.0)

> Pittsburgh, Pennsylvania, is home of the Pittsburgh Steelers (Weight 4.0 + 1.0)

Now compare these two sentences to the following: "Pennsylvania has a number of cities involved in manufacturing, shipping, and trade, including Pittsburgh, Erie, Allentown, Scranton, Reading, and Philadelphia." Based on the weighting we have given for the other two sentences, this sentence has a weight of 1.0 + 1.0 for the single instances of "Pittsburgh" and "Pennsylvania" in it. But, the other two sentences are each 8 words long, and

Table 2.1 Mean (μ), standard deviation (σ), entropy (e), and coefficient of variance (COV = σ/μ) for the decile distributions in a document of the two words A, with decile counts of {5, 4, 3, 7, 8, 4, 4, 5, 7, 3}, and B, with decile counts of {0, 1, 0, 14, 21, 9, 2, 2, 0, 1}. The maximum possible e is 3.32, which is log base 2 of 10 (since the distribution has 10 elements).

Word	Mean μ	Standard deviation σ	Entropy e	COV
A	5.0	1.67	3.24	0.33
B	5.0	6.91	2.08	1.38

this sentence is 20 words long. It seems logical to conclude that Pittsburgh and Pennsylvania have a smaller overall role in this sentence. If we divide by the length of the sentence, then the original two sentences have weights of 0.25 and 0.625, respectively, while the third sentence has a weight of 0.10. We will go no further on this type of normalization since it starts to look like the next method, to which we now turn our attention.

(2) **Word frequency** is the second method for simple weighting for extractive summarization. At its simplest, word frequency is basically the same as word scoring, wherein the score for selecting an element of text is simply the sum of the occurrences of the words that are important to the user (e.g., in the search query, in the set of indexed terms, etc.). However, weighting by terms can be a non-linear (e.g., geometric, power, exponential, logarithmic, etc.) function of the prevalence of the word in a phrase, a sentence, a paragraph, or other text element, as shown above. Perhaps more importantly, it can be referential to a given set of context. If, for example, we have two words, each of which occurs 50 times in a large document but with the following occurrences in the deciles of the book:

{5, 4, 3, 7, 8, 4, 4, 5, 7, 3} (Word A)

{0, 1, 0, 14, 21, 9, 2, 2, 0, 1} (Word B)

From an information standpoint, we note that Word A is much more randomly distributed throughout the document; that is, it has high entropy. Word B is centered on the 4th, 5th, and 6th deciles of the document. As a consequence, we obtain widely different statistics about these two words, even though they both occur 50 times in a single document. The mean, standard deviation, entropy, and coefficient of variance of these two 10-element distributions are shown in Table 2.1.

Using Table 2.1 as a guide, then, we should be able to weight the word frequency by the distribution. In general, the higher the entropy of the distribution, the more dispersed the term is throughout the document, and, thus, the less valuable it will tend to be as a term for locating important extractive summarization phrases. The lower the entropy, all other factors being equal, the more concentrated the word will be in a specific section of the document, and, thus, the more likely that specific section is to be one that produces a phrase worth belonging to the extractive summary. Thus, we can adjust the weight of a specific word for extraction summarization value by multiplying it by the inverse of its entropy. Word A, thus, goes from word frequency weight of 50.0 to one of 50.0/3.24 = 15.4, while Word B goes from word frequency weight of 50.0 to one of 50.0/2.08 = 24.0.

Similarly, the variability in the deciles can be used for word frequency weight adjustment. The higher the variability, in general, the less uniform the distribution and the more a specific word is concentrated in a particular decile or small number of deciles. Thus, in this case, we can adjust the weight of a specific word for extraction summarization value by multiplying it by the variance. Word A, thus, goes from word frequency weight of 50.0 to one of 50.0 × 1.67 = 83.5, while Word B goes from word frequency weight of 50.0 to one of 50.0 × 6.91 = 345.5. If we then multiply by the inverse of the entropy, the weight for Word A is 83.5/3.24 = 25.8, and the weight of Word B is 345.5/2.08 = 166.1. Using these two factors simultaneously has a powerful effect on the relative weight of two terms having the same word frequency in the document. We could also use the COV (Table 2.1) to further affect the relative weights. In every case, however, we would need to set a factor based on empirical data, resulting in Equation (2.1):

$$\text{Word Frequency Weight} = \frac{(N_{\text{words}}) \times \text{COV} \times \sigma}{e}. \qquad (2.1)$$

One additional source of weighting for word frequency will be included in this section. In linguistics, there are many ways in which more frequent occurrence of a given word is hidden through literary techniques. The easiest ones to describe are synonyms, where one word that means the same thing as another word is used in place of it. For example, in one sentence, a "dish" is used, and in the next, a "plate." A "car" drives around one sentence and enters the next street as an "automobile." In this case, we would, on the surface, simply sum the synonyms and add them to the count of the particular word; that is, we would increment N_{words} in Equation (2.1). However, some words are more synonymic than others. For example, "vehicle" is a synonym for "car" but also for "bicycle," "motorcycle," and "truck." In this case, the degree of similarity, a value between 0 and 1, should be accounted for. So, an effective word frequency for each synonym is computed based on Equation (2.2). In Equation (2.2), N_{synonyms} is the number of times the synonym occurs, and Similarity$_{\text{synonym}}$ is its degree of similarity to the particular word that we are weighting.

$$\text{Effective Word Weight} = \text{Similarity}_{\text{synonym}} \times (N_{\text{synonyms}}). \qquad (2.2)$$

A similar approach is used to count the contribution of three types of figures of speech, as well. The first is metonymy, which changes (meta) of name (nym). A metonym is when a substitute name, term, or expression is used in place of a closely related name, term, or expression. Examples of metonymy are "the bottle" for "strong drink," "Wall Street" for the "financial industry," and "the idiot" for "Uncle Greg." OK, maybe I should not have listed the last one, but the point is clear. These terms are as closely related as traditional synonyms. The second figure of speech is the synecdoche, which is where a part of a thing represents the whole, or vice versa. Examples of synecdoches are "three sail" for "three ships," an "Einstein" for any/all brilliant people, and "mouth" instead of "children" in a large family where the parents have many "mouths" to feed. The third figure of speech of interest here is the kenning, which is a situation in which we substitute a phrase for another phrase, usually as a figurative expression replacing a noun. Kennings are an often sophisticated literary technique used, for example, considerably in Sage Age Icelandic literature, where a "wave traveler" might represent a "boat." More modern kennings include a "book worm" for an "avid reader" and a "meat bag" for a human that is working side by side with a robot. Once we have determined the figure of speech (metonym, synecdoche, kenning, etc.) and the actual word in the text element that it represents, we can then "add" the weight of the number of occurrences of the figures of speech to the weight of the word itself. Sounds straightforward enough!

However, mapping these figures of speech to their appropriate "anchor" word is not always easy and, as for the synonyms above, may not always be a 100% match. For example, if we read a passage about Amerigo Vespucci setting forth with "three sail" in 1501, then how do we map this to the names of each ship? Do we assign a weight of 1/3 to each, or 0.5 to the flagship and 0.25 to the other two ships? The fact of the matter is, this is an ambiguous situation, and the best way to assign partial weighting is really most properly determined through training and validation of the system. Different functional goals for the system are almost assured to result in differences in how the partial, or "effective," weighting is to be assigned. For example, suppose the name of the flagship is "Sevilla," and the other two ships are the "Cadiz" and the "Malaga" (I do not know the real names of Vespucci's ships, so these are a good proxy set of names given where he retired). If the end goal of the summarization is to create a summarized history of the famous voyages of "discovery" in the late 15th and early 16th century, then it may be appropriate to assign the 1.0 sum weight for "three sails" as {Sevilla, 0.5}, {Cadiz, 0.25}, and {Malaga, 0.25}. If, however, the end goal of the summarization is to catalog all of the ships used in early exploration, then it may be more appropriate to assign the 1.0 sum weight for "three sails" as {Sevilla, 0.333}, {Cadiz, 0.333}, and {Malaga, 0.333}.

Whatever we decide in the assignment of effective weighting for all of the synonyms and figures of speech, we use Equation (2.3) for the final word frequency weight assignment, where the $N_{\text{effectivewords}}$ is a multiple between 0.0 and 1.0 of the sum of all appropriate synonyms and figures of speech referring to each word.

$$\text{Word Frequency Weight} = \frac{(N_{\text{words}} + N_{\text{effective words}}) \times \text{COV} \times \sigma}{e}.$$

(2.3)

With this consideration of word scoring and word frequency completed, we can now turn to the relative frequency of words, focusing on TF*IDF.

(3) **TF*IDF** is the third important means of weighting individual terms to prepare them for extractive summarization. The TF is the term frequency, which was introduced above as the word frequency. Here, a term can be a compound noun, like "São Paolo" or "Kennedy Space Center," or even a longer phrase, such as "We are such stuff as dreams are made on, and our little life is rounded with a sleep," a quote from Shakespeare's *The Tempest*, which, like the *Hamlet* quotes above, are otherwise very unlikely to show up in a summarization. However, once we aggregate the entire string of words into a single element, the relative frequency in any corpus of documents drops considerably from the frequency of any of the individual words.

The IDF term is the inverse document frequency, which simply means the inverse of the document frequency, or the baseline level of word occurrences in the entire corpus (or another referent set). Using the example above, "We are such stuff as dreams are made on, and our little life is rounded with a sleep" occurs once in *The Tempest* and once in the entire set of Shakespeare's plays. Thus, TF = 1.0 and DF = 1/37 = 0.027; since the expression occurs once in the set of 37 Shakespeare plays, it occurs 0.027 times per play. Thus, the simplest form of TF*IDF for this expression in *The Tempest* is $1.0 \times (1/0.027) = 37.0$, which means, of course, that it occurs 37 times as commonly in *The Tempest* as in Shakespeare's plays in general. There are at least 112 forms of TF*IDF, as will be described later in this chapter so that in this section, we focus on the root form of simply taking the multiple of the occurrences in a given text element with the inverse of the mean number of occurrences in all other documents from which we can choose.

Let us apply this simplest of TF*IDF approaches to the four sentences below, where we assume each sentence is a (very short) document:

(1) Pittsburgh, Pennsylvania, is where the Ohio River starts
(2) Pittsburgh, Pennsylvania, is home of the Pittsburgh Steelers
(3) To be or not to be, that is the question
(4) We are such stuff as dreams are made on, and our little life is rounded with a sleep

At the word level, Pittsburgh occurs 3.0 times (assuming that the word "home" is not somehow mapped to "Pittsburgh" in number 2). The DF = 3.0/4.0 = 0.75, and the IDF = 1.333. Thus, "Pittsburgh" occurs a mean of 0.75 times in a document. For the four documents, the TF*IDF of "Pittsburgh" is

(1) TF*IDF(Pittsburgh) = 1.0 × 1.333 = 1.333.
(2) TF*IDF(Pittsburgh) = 2.0 × 1.333 = 2.667.
(3) TF*IDF(Pittsburgh) = 0.0 × 1.333 = 0.0.
(4) TF*IDF(Pittsburgh) = 0.0 × 1.333 = 0.0.

We can see a wide range of TF*IDF values for "Pittsburgh" in these four documents. If a search on "Pittsburgh" were performed on this small set of documents, the search order would be Document 2, then Document 1, with Documents 3 and 4 having no match. Suppose instead we wished to use a common word like "is" to see the search order. The word "is" occurs exactly 1.0 times in each of the four sentences, and, thus, for each of the four sentences, TF*IDF(is) = 1.0.

From this section, we see the importance of TF*IDF for identifying where terms are focused in a larger corpus. As our simple example here illustrates, TF*IDF can be used to highlight sections within larger bodies of text that should be "jumped" to when a positive search result is returned. For example, in the document for which the decile occurrence histogram of Word B was {0, 1, 0, 14, 21, 9, 2, 2, 0, 1}, when it is returned as a positive search, the cursor should most likely jump right to the fourth decile since the following three deciles are where Word B is most highly concentrated. Combining these approaches allows us to be far more accurate in searches (or waste less time finding the best location within a document that is returned from the search).

(4) **Upper case** is the fourth approach for use to differentially weight terms in preparation for extractive summarization. There are two primary forms of first-letter upper case, or capitalization, at least in English. The first is the capitalization at the beginning of each sentence, which is effectively a non-starter for differential weighting of sentences since, by definition, each sentence will have exactly one of these. As shown in the example of TF*IDF for the word "is" in the prior section, this does not lead to differentiation among the sentences. Capitalizing the first word after a colon or the first word in a quote when it begins a sentence within the quote essentially devolves to the same thing as the "first word in a sentence" rule and, so, is not of interest to us. However, the second circumstance in which words are capitalized is of interest. Proper nouns, including names of specific locations, organizations, people, or things, must be capitalized. From Texas to the Texas Rangers, and from George Washington to the Washington Monument, proper nouns are often words of relatively high importance in a

text passage. Anything that becomes something more specific, whether it be Middle Earth, the Middle Ages, or Middle English, needs capitalization. With this capitalization, presumably, come privileges and prestige. In the world of extractive summarization, this means weighting words with capitalization above 1.0. Our rule might be along the lines of the following:

(a) For every word or compound word (e.g., "Salt Lake City" counts only as a single capitalization since it properly represents only a single place name), assign a weighting of 1.5.

(b) If the word occurs at the beginning of the sentence, then reassign a weight of 1.0.

(c) If the first word in the sentence would otherwise be capitalized within the sentence, then assign it a weight of 1.6.

Rule (c) provides a slight bonus for a proper noun at the beginning of the sentence, which seems reasonable as it is more likely to be the subject of the sentence. For example, compare the now familiar sentence, "Pittsburgh, Pennsylvania, is where the Ohio River starts," with the new sentence, "I've never visited Pittsburgh, even though I've been to Akron." Weighting "Pittsburgh" slightly higher in the first sentence seems justified since it is the subject of that sentence and the indirect object of the second sentence. The actual relative weights in (a) and (c), of course, may not be 1.5 and 1.6 but are settings that can be readily set in the validation stage of the summarizer's development. For example, these numbers may end up being 1.35 and 1.55, respectively. The values may change, but their order is not likely to.

This handles capitalization, but what about additional upper case? It might seem like additional upper case should count for more since there are more capitalizations. However, the UN is the United Nations and NYC is New York City, and there is no reason why the shortened form should count for more. Based on our discussion of synonymy and figures of speech, in fact, the absolute upper bounds for the UN and NYC are the weights given to the original terms. Since they are direct synonyms, we can, therefore, claim that **an acronym has the same weight as the longer expression it represents**. But what about other forms of multiple capitalization? For example, what of onomatopoeic expressions like "OW!" and "WHAM!" and "SHAZAM!"? Should they be weighted more heavily or not? We argue that cases such as these are treated the same as boldface, italics, underlining, enclosing in two asterisks, or any other form of emphasis. For words like this, the capitalization is integral to the emphasis, not to the same consideration of importance for proper nouns. In our experience, weighting given to emphasis is often considerably less than that given to capitalization. A relative weight of 1.05 to 1.1 generally suffices.

(5) The **Proper noun** is the fifth consideration for differential word weighting in extractive summarization. The capitalization aspect of the proper noun is addressed in the previous section. In this section, the additional value of the proper noun as an indexing, or tagging, item is considered. As noted above, proper nouns are **special** names, locations, organizations, and generally any special thing. Thinking of manufacturing as an example, common nouns are akin to mass production, while proper nouns are akin to custom manufacturing. Proper nouns are special instances of common nouns, individual incarnations. A boy is a boy, but Billy Bob is special. A memorial is solemn, but the Taj Mahal is especially solemn. A month covers a cycle of the moon, but June is the month where the moon shows the least. A country is a homeland, but China may well be your particular homeland. Are all proper nouns created equal? Maybe a person is more important than a place, and a country more important than a book or film. It is safe to say that the different types of proper nouns are almost assuredly deserving of unequal weighting, depending on the particular linguistic application. For example, if you are providing some type of summarization of the content on IMDB.com, a well-known movie, television, video, video game, and streaming content database, you are almost certainly better off weighting the titles of content and personal names more highly than specific locations and country names. The relative additional weighting for proper nouns is, therefore, relative to other proper nouns and is multiplied with the weighting for capitalization, which may already be $1.5-1.6$ as pointed out in the previous section. The additional relative weighting is, therefore, relatively close to 1.0. Often, using $1.01-1.09$ for the relative weighting of different types of proper nouns will suffice. Different types of proper nouns can be distinguished using lists (taxonomy-driven) or determined with machine learning approaches.

(6) **Word co-occurrences** provide the sixth consideration for the differential weighting of words in anticipation of phrase-selective, that is, extractive, summarization. Co-occurrence, to a linguist, is the above-random probability that two or more words, expressions, or phrases will occur in close proximity in a text element. In some cases, this means the terms have semantic proximity or that one is an idiomatic expression for the other. The next section will address dialect. Here, we are concerned with the statistical analysis of word frequencies and other patterns of occurrences, including coincidence (occurring together in some documents with higher frequency, but overall covering a normal distribution of co-occurrence) and collocations (occurring close together in a document with statistically relevant predictability). For the purposes of summarization, the mathematics in this section are relatively straightforward. For each pair of words, expressions, or phrases in a document, you compute the mean closest

distance between them and every other word, expression, or phrase in the document. If there are W words in a document, then there are $W(W-1)/2$ such calculations to make. For example, suppose the two words you are looking at for possible co-occurrence are "snake" and "dragon." In a document 8000 words long, there are four and five instances of "snake" and "dragon," respectively, occurring as follows:

(1) Snake at word locations 810, 1415, 4918, and 6317
(2) Dragon at word locations 844, 1377, 1436, 5043, and 5096

From these, we can see the mean of the minimum distances between "snake" and "dragon" is the mean of 34, 21, 125, and 1221, which is 350. This is well below the random mean minimum distance between an instance of "snake" and "dragon" for such a scenario, which is 762, as computed in the R code of Figure 2.1 based on 100,000 random instances. The ratio of 350/762 is 0.459, which is well under the expected median value of 1.0 for any two terms. The pairings with the highest level of co-occurrence will be defined as those with the lowest ratio of actual to random distancing, as determined by this process. This is actually a computation of collocation more than co-occurrence and, so, is designated the CollocationRatio(A, B) for Term A to Term B. This CollocationRatio is defined in Equation (2.4).

$$\text{CollocationRatio}(A, B)$$
$$= \frac{\text{Mean Minimum Distance from Term } A \text{ to a Term } B}{\text{Expected Value of Numerator}}. \quad (2.4)$$

In Equation (2.4), CollocationRatio(snake, dragon) = 350/762 = 0.459, as mentioned above.

Note that CollocationRatio(A,B) is usually not the same as CollocationRatio(B,A) since these two terms rarely have the same number of occurrences, and even if they do, it does not guarantee that minimum distance "pairings" are 1:1. For "dragon" to "snake," for example, the minimum distances are 34, 38, 21, 125, and 178, for a mean of 79, well less than the 350 from "snake" to "dragon." Here, for "dragon," the ratio is much lower than that for "snake." Running the code in Figure 2.1 with the j and k for() loops reversed, we find after 100,000 iterations that a mean minimum distance from each instance of "dragon" to the nearest instance of "snake" is 1168, meaning that CollocationRatio(dragon, snake) = 79/1168 = 0.068, far less than 1.0 and also far less than the 0.459 value for the converse CollocationRatio(snake, dragon).

Moving past the specific example to the general case, once these ratios (Equation (2.4)) are computed for all terms, expressions, and phrases of interest, we can then weight them relative to the inverse of their key co-occurrences. Since the mean value of CollocationRatio() across a

```
random()
{
        N = 100,000
        i = 0
        total_distance = 0
        for(i in seq(from = 1, to = N, by=1))
        {
                snake = runif(4, 1, 8000)
                dragon = runif(5, 1, 8000)
                min_distance = 8000
                for(j in seq(from = 1, to = 4, by = 1))
                {
                        min_distance = 8000
                        for(k in seq(from = 1, to = 5, by = 1))
                        {
                                if( abs(snake[j]-dragon[k]) < min_distance )
                                        min_distance = abs(snake[j]-dragon[k])
                        }
                        total_distance = total_distance + min_distance
                }
        }
        mean_distance = total_distance/(4*N)
        print(total_distance)
        print("mean_distance")
        print(mean_distance)
}
```

Figure 2.1 Simple *R* code to calculate the random distances between four instances of "snake" and five instances of "dragon" in a document that is 8000 words long. The output of the print() statements are 304,694,181, "mean_distance," and 761.7355. Please see text for details.

(presumably large) set of text elements should be close to 1.0, using the inverse of the collocation ratios and taking their mean provides a good estimate for the relative co-occurrence of two text elements. In our case, CollocationRatio() values for the pair "snake" and "dragon" are 2.18 and 14.71, the mean of which is 8.45. Thus, these two terms are co-occurring roughly 8.5 times more than we would anticipate by chance (never mind what that actually means since it is a relative measure). We do not, of course, assign a weighting of 8.5 to this pair; instead, we assign weightings based on bands of relative co-occurrence to the expected value of 1.0. The range of weightings assigned is generally rather small here for the data sets that we have used them on, and an example is given here:

(a) If **mean(CollocationRatio())** **of pair (*A*, *B*)** > 5, then assign weighting 1.1.

(b) If 2 < **mean(CollocationRatio())** of **pair** (*A*, *B*) < 5, then assign weighting 1.05.
(c) If 0.5 < **mean(CollocationRatio())** of **pair** (*A*, *B*) < 2, then assign weighting 1.0.
(d) If 0.2 < **mean(CollocationRatio())** of **pair** (*A*, *B*) < 0.5, then assign weighting 0.95.
(e) If **mean(CollocationRatio())** of **pair** (*A,B*) < 0.2, then assign weighting 0.90.

Your results will vary, but as with each of the other factors we are considering for weighting, it can be tested relatively independently of the other factors. Our advice is to err on the side of keeping the weightings close to 1.0 for this factor since the values for CollocationRatio() will tend to be more variable for terms with less overall numbers of occurrences, and these terms will tend to have higher TF*IDF and other scores already. You do not need a lot of differentiation here; in fact, co-occurrence calculations may be better suited to finding synonymic and figure of speech (metonym, synecdoche, kenning, etc.) rules among terms than for introducing a new set of weightings. With that, we move on to the related topic of lexical similarity.

(7) The next factor worth considering for text element weighting for extractive summaries is **lexical similarity**. While it can be uncovered in some cases by co-occurrence, as mentioned above, lexical similarity is a potentially wide set of measures relating the degree of similarity when comparing two word elements. At the broadest level, lexical similarity is used to determine if two languages have similar word sets. This ties to the term "lexicon" and is effectively comparing the breadth of one language versus another. While an admirable text analytic to compute, it is, nevertheless, not the type of lexical similarity that we are concerned with here.

Instead, we are concerned in this section with the lexical similarity of the different potential elements of the extractive summary with the text used to interrogate these elements. Here, lexical similarity tries to match words and expressions in pairings between the two texts. The simplest means of performing lexical similarity is, perhaps, the vector space model (VSM). The VSM provides a useful similarity that ranges from 0.0 (no words shared between two text elements) and 1.0 (e.g., the exact same document compared to itself or two documents with the exact same word histograms). The "vector space" of the VSM is an N-dimensional space in which the occurrences of each of N terms in the text element are plotted along each of N axes for the text elements. Suppose we have two text elements that we will call document a and document b. The vector $\vec{a}$ is the line from origin to the term set for document a, while the vector $\vec{b}$ is the line from origin to the term set for query b. The dot product of $\vec{a}$ and $\vec{b}$, or $\vec{a} \bullet \vec{b}$, is given by Equation (2.5):

$$\vec{a} \bullet \vec{b} = \sum_{w=1}^{N} a_w b_w. \tag{2.5}$$

Using this definition, the cosine between the query and the document is readily computed using Equation (2.6):

$$\cos(\vec{a}, \vec{b}) = \frac{\vec{a} \bullet \vec{b}}{|\vec{a}| \, |\vec{b}|} = \frac{\sum_{w=1}^{N} a_w b_w}{\sqrt{\sum_{w=1}^{N} a_w^2} \sqrt{\sum_{w=1}^{N} b_w^2}}. \tag{2.6}$$

Let us apply this to three familiar sentences that we have used throughout this chapter, where the percent of the sentence that is each word is given in parentheses following the word:

(1) Pittsburgh (0.125), Pennsylvania (0.125), is (0.125) where (0.125) the (0.125) Ohio (0.125) River (0.125) starts (0.125).
(2) Pittsburgh (0.25), Pennsylvania (0.125), is (0.125) home (0.125) of (0.125) the (0.125) Pittsburgh (0.25) Steelers (0.125).
(3) To (0.2) be (0.2) or (0.1) not (0.1) to (0.2) be (0.2), that (0.1) is (0.1) the (0.1) question (0.1).

The reader can readily compute the cosines between these three sets as

(1) cosine(1, 2) = 0.559
(2) cosine(1, 3) = 0.189
(3) cosine(2, 3) = 0.0.169

Clearly, the first two sentences are more similar. What may be of interest is that sentences 1 and 3 are more similar than sentences 2 and 3, even though each pairing shares the same two words with the exact same weights in common. Thus, the denominator in Equation (2.6) is the same for both cosine(1, 3) and cosine(2, 3). It is just that the numerator term is larger for sentence 2 than sentence 1 because of the double occurrence of the word "Pittsburgh." The cosine measure is also known as the *normalized correlation coefficient*.

In between cosine similarity and lexicon similarity, lexical similarity can also be used to determine the key terms in a dialect. This is a form of meta-summary but is still an extractive mechanism. Dialect idioms and expressions can be determined by comparing the dictionaries of terms (used for the VSM) and assessing which terms are highly different in occurrence rate in the two dialects for documents covering the same subject matter.

(8) **Sentence scoring** is another consideration in the preparation of text for summarization. In a sense, all of the above methods involve, in the

deployment phase, some form of sentence scoring since the sum of all the weights for all of the words in a sentence combines to give a score that allows sentences to be selected for relevance in the final summarization. However, by the term "sentence scoring" here, we mean an independent method of scoring the sentence based on an authority, or expert. This expert provides a broad set of domain-specific terms that can be used to score the sentences. Among the more common ways to achieve sentence scoring is to have the cue words entered into a database by practitioners, or experts, in the fields of interest. In many ways, this is analogous to a specific set of search queries. However, the taxonomy can be defined by an authority in a field of expertise distinct from the text analyst without any loss of accuracy. Once the set of cue words are defined, above methods such as word frequency and co-occurrence can be used to assess the relevance of a text element to each of the classes defined in the taxonomy.

Applying this to the following three now-familiar sentences, we use the example of three simple taxonomies to determine which sentence is most likely to belong to an extractive summary of a document containing all three sentences. The three sentences are as follows:

(A) Pittsburgh, Pennsylvania, is where the Ohio River starts.
(B) Pittsburgh, Pennsylvania, is home of the Pittsburgh Steelers.
(C) Pennsylvania has a number of cities involved in manufacturing, shipping, and trade, including Pittsburgh, Erie, Allentown, Scranton, Reading, and Philadelphia.

The three taxonomies are as follows:

(1) [Cities in Pennsylvania]
(2) [States in the US]
(3) [NFL cities]

These categories really need no elaboration, so, instead, we look at the matches between each of the three sentences and the three categories. The matches are collected in Table 2.2. The data in the table can be looked at from two perspectives: (1) identifying which of the summaries a sentence would best be assigned to, and (2) identifying which of the sentences should best be included in a summary. For the first consideration, Sentence (A) matches best to "States in the US" and Sentence (C) matches best to "Cities in Pennsylvania." Sentence (B) is a little less obvious since it has a score of 2 for both "Cities in Pennsylvania" and "NFL cities." And, no matter what the term weighting approach is, the weight will be the same since the score is for the dual occurrence of "Pittsburgh." Thus, without further context, we are equally likely to assign Sentence (B) to either of these two categories (or both, if allowed).

Table 2.2 Matches for each of the three sentences for each of the three categories. Numbers represent the absolute number of terms matching.

Sentence	Cities in Pennsylvania	States in the US	NFL cities
(A)	1	2	1
(B)	2	1	2
(C)	6	1	2

Table 2.3 Matches, normalized by length of sentence, for each of the three sentences for each of the three categories. Numbers represent the density (percent of all terms) of terms matching.

Sentence	Cities in Pennsylvania	States in the US	NFL cities
(A)	0.125	0.250	0.125
(B)	0.250	0.125	0.250
(C)	0.300	0.050	0.100

However, summarization works in the inverse of the method just described. For the summarization of "Cities in Pennsylvania," Sentence (C) is selected based on the number of matching terms. Even if the matching terms are normalized by the overall number of words in each sentence (Table 2.3), Sentence (C) is still chosen first to be part of the summary on the basis of the largest value (0.300). Similarly, Sentence (A) is chosen first for the "States in the US" summary on the basis of its highest value among the sentences in both Tables 2.2 and 2.3.

Deciding which sentence to choose first for the "NFL cities" is not possible using only Table 2.2 since both Sentences (B) and (C) share two matches and the same score. Interestingly, if we would weight terms occurring more than once more heavily (see section on Word Scoring above), then we would select Sentence (B). Also, when we normalize the scores for the length of the sentence (i.e., the density of matching words, as captured in Table 2.3), we select Sentence (B). This is in spite of the fact that Sentence (C) actually has two NFL cities (Pittsburgh and Philadelphia) in its word list, and Sentence (B) only one (Pittsburgh). However, were we to limit a summary to 200 words, then our expected value of unique NFL cities for 200 words of Sentence (B)-like content is 25, and the expected value of unique NFL cities for 200 words of Sentence (C)-like content is 20.

Other forms of cue word sets include keywords mined from all of the abstracts belonging to a topical journal, indexing performed by library scientists, meta-data associated with electronic files, and more. The commonality is that there is a specialized terminology that is a subset of the general language. For any such subset, selecting classes to assign the sentences, and the converse assigning sentences to summaries, is readily accomplished by the means of weighting described in this section and chapter.

(9) **Cue phrases** are text expressions that give discourse clues because of the parts of speech they occupy. Common expressions used as cues include "anyway," "by the way," finally," "first, second, third," "for example," "furthermore," "hence," "in summary," "incidentally," "lastly," "moreover," "now," "on the other hand," "then," "therefore," and "thus" [Hoar98]. Each of these expressions give a hint that what follows them is rather significant, and, thus, the phrase following any of these expressions might be weighted above the normal. Again, having reached our ninth factor that might boost expressions for selection by an extractive summarization engine, it may come as little surprise that this value will generally be only slightly above 1.0. Trying the values {1.00, 1.01, 1.02, ..., 1.10} and selecting the one in this range that optimizes the summarization performance on ground truthed text sets is almost certainly sufficient to address cue phrases.

(10) **Sentence inclusion of numerical data** is another minor indicator of potential value for extractive summarization. As an example, consider the two following sentences:

(A) Pennsylvania has a number of cities involved in manufacturing, shipping, and trade, including Pittsburgh, Erie, Allentown, Scranton, Reading, and Philadelphia.
(B) Pennsylvania has at least 6 major cities involved in manufacturing, shipping, and trade: Pittsburgh, Erie, Allentown, Scranton, Reading, and Philadelphia.
    Each sentence has 20 words, with the primary difference being Sentence (A) says the word "number" whilst Sentence (B) actually types the number "6." If numerical data in a sentence has summarization "affinity," certainly it cannot be very much since these sentences are basically identical. Therefore, under most circumstances, we apply a multiplication factor of 1.0 to sentences with numbers in them (that is, no effect). However, this is subject to a threshold. Suppose, for example, we find Sentence (C) to be the following:
(C) There are three Pennsylvanian cities with more than 100,000 inhabitants: Philadelphia, with 1,567,000; Pittsburgh, with 304,000; and Allentown, with 120,000.

Here, the mention of "three" has no impact, but the other four numbers comprise 20% of the terms in the sentence and include 25 digits. We may therefore craft additions to sentence weighting based on a set of rules of the count of numerical terms (here 4) and the total number of digits (here 25), such as:

(A) Set Numerical_Weight = 1.00.

(B) For all numerical terms with 3 or less digits, assign 0.00 additional weight.
(C) For all numerical terms with 4 or more digitals, assign 0.01 additional weight.
(D) If the sentence is only numerical in nature, reset weight to 1.00 (may be improperly parsed or just be an entry in a table, in either case of which extra weight should not be assigned).

As is clear from this algorithm, the relative impact of numerical weighting on extractive summarization is relatively modest.

(11) **Sentence length** is another factor to be considered in the generation of sentence weighting for extractive summarization. In general, longer sentences contain more keywords and key phrases, so their weighting naturally increases. A sentence $3\times$ as long as another sentence is expected to have three times the keywords (words which have rareness values above 1.0, for example), in addition to more upper case words, more pronouns, and more co-occurrences, to name just a few. Thus, longer sentences already benefit from most of the weighting approaches listed so far. Without corrective balance, then, longer sentences will always be chosen for summaries. "What's the problem?" you may ask, since if they are more likely to contain salient terms and phrases, should they not comprise the summary?

The problem is that a summary is meant to compress a document. Selecting longer sentences works against compression. If we have a Pareto-like situation wherein 20% of the sentences have 80% of the length of the document, and summarization is performed by selecting 10% of the sentences, then we can readily understand this summary might comprise 40% of the document in length. That is not a particularly compressed document. Therefore, in general, sentence weighting is proportional to the inverse of sentence length. This is true, however, only for sentences above a threshold value of length. We also do not want to select only short sentences, like "Wham!" or "Oh no!" or "Stop!" since none of these conveys much information helpful to understanding the gist of a document. We may, as a consequence, use a sentence length rule along the lines of:

(A) For a sentence < 10 words, sentence length weight $= 1.0$.
(B) For a sentence > 10 words, sentence weight $= 1.0 - 0.01 \times$ (length in words $- 10$).

If a sentence is 20 words long, then, from (B), its weight is $1.0 - 0.01 \times (20 - 10) = 0.90$.

(12) **Sentence position** is another factor to consider for sentence weighting. Several possible interpretations are possible. One we have used in our research is that the positional value of a sentence is given in terms of ranking the sentences from 1 to N, where N is the number of sentences

in the document. Rank 1 is assigned to the first sentence, rank 2 to the last sentence, rank 3 to the second sentence, rank 4 to the second to last sentence, and so on. Then, the weighting of the sentence varies from 1.0 to 1.0 + Sentence_Position_Factor, where Sentence_Position_Factor varies from 0.1 to 0.5, depending on the types of documents being analyzed. Sentence_Position_Factor is typically higher for shorter documents since they often have attention-getting first sentences and memorable last sentences. The rank influences Sentence_Position_Factor as defined in Equation (2.7):

$$\text{Weighting} = 1.0 + \text{Sentence\_Position\_Factor } x \left(\frac{N - \text{Rank} + 1}{N} \right).$$
(2.7)

In Equation (2.7), when Rank = 1, the weighting is 1.0 + Sentence_Position_Factor, and when Rank = N, the weighting is 1.0 + Sentence_Position_Factor $(1/N)$, which for large values of N is approximately 1.0. Note that this approach can also use a non-linear relationship between the maximum weighting and the minimum. For example, the Rank-dependent part of Equation (2.7) can be given a power factor, P, as shown in Equation (2.8). Some preliminary work we have done with this metric indicates that $P = 2$ is not a misguided choice.

$$\text{Weighting} = 1.0 + \text{Sentence}_{\text{Position}_{\text{Factor}}} x \left(\frac{N - \text{Rank} + 1}{N} \right)^{P}.$$
(2.8)

Overall, Sentence_Position_Factor is an important consideration for extractive summarization, and one that we will come back to when revisiting meta-algorithmics and summarization. It is certainly safe to say that many writers wish to start and end strongly, and so the first and last sentences are usually more likely to be part of an extractive summarization than the sentences in between.

(13) **Sentence centrality** is our next factor to consider. The centrality of a sentence is also open to various interpretations. Our perspective on sentence centrality is that the most central sentences are those that connect best to the rest of the sentences in a document. In a sense, this is analogous to the Page Rank algorithm [Brin98], which ranks pages based on their connectedness as well as the score of the pages connecting to them. Here, our sentence centrality is based on its connectedness to other "high ranked" sentences. To be a high-ranked sentence, there are several options, including at minimum the following:

(A) Mean word weighting in sentence is x (where typically x is in the range of $0.3-1.0$) standard deviations above the mean of all sentences.

(B) Mean word weighting in sentence is greater than some threshold, e.g., 1.5.
(C) Sentence is among the *P* percent highest-ranked sentences, where *P* is typically in the range of 0.05–0.20, depending on size and connectedness of the set of sentences.

Interestingly, these methods of sentence centrality are a way of finding sentences that are important for connecting the sections of a document together, which is a different way of looking at extractive summarization than simple relative weighting. The sentences best connected to the other sentences, we assume, provide a better overall flavor of the document than sentences that might be weighted heavily but are disconnected from the rest of the sentences in a document. Thus, the weighting applied to sentences for centrality, like those for word frequencies, TF*IDF, and sentence position, can be relatively high. The most central sentence might receive a weighting factor several times that of no centrality, i.e., 3.0 compared to 1.0.

(14) **Sentence resemblance to the title** is a simple, but often effective, weighting approach for extractive summarization. This is effectively a simple computation of lexical similarity between each sentence *S* and the title *T*, with the vector $\vec{S}$ being the line from origin to the term set for sentence *S*, while the vector $\vec{T}$ is the line from origin to the term set for the title *T*. The dot product of $\vec{S}$ and $\vec{T}$, or $\vec{S} \bullet \vec{T}$, is given by Equation (2.9):

$$\vec{S} \bullet \vec{T} = \sum_{w=1}^{N} S_w T_w. \tag{2.9}$$

Using this definition, the cosine between the sentence and the title is

$$\cos(\vec{S}, \vec{T}) = \frac{\vec{S} \bullet \vec{T}}{|\vec{S}||\vec{T}|} = \frac{\sum_{w=1}^{N} S_w T_w}{\sqrt{\sum_{w=1}^{N} S_w^2} \sqrt{\sum_{w=1}^{N} T_w^2}}. \tag{2.10}$$

A simple means of weighting a sentence for similarity to the document is to assign it a weight as defined by Equation (2.11):

$$\text{Title\_Similarity\_Weighting} = 1.0 + K\cos(\vec{S}, \vec{T}) \tag{2.11}$$

where the coefficient, *K*, is determined experimentally. Typically, setting *K* between 0.5 and 2.0 provides approximately the correct weighting for similarity to the title.

(15) **Graph scoring** is the final factor we consider here for extractive summarization. There are many possibilities for graph scoring, but, essentially, they are performing matching between the topics of a

given document and the topics of interest to the person requesting the summarization. As such, graph scoring has a bit of a different flavor to it than the more generic factors described in the rest of this section. Graph scoring reaches outside of the given document (i.e., beyond the title) to the "template" document to which it is matched (or not matched). Graph scoring can be done at the sentence diagram level for cases in which the template is a short document (e.g., a sentence or search query) or at the document level for comparing, among other possibilities, similarity in word histograms, rare word sequences, etc. Aggregate similarity is when multiple graph scoring approaches are used at the same time.

Having considered in some depth these 15 factors in extractive summarization, we now describe how they are used together. Fortunately, this is uncomplicated. You have probably noticed that all of the weights described are centered around 1.0. It is so that all 15 of them can be multiplied together to get the weighting for any individual word, and then the sentence score is simply the sum of these individual multiples. This is shown by Equation (2.12).

$$\text{Total}_{\text{Sentence}_{\text{Score}}} = \sum_{i=1}^{N} \prod_{k=1}^{15} \text{WF}_i(k). \qquad (2.12)$$

Here, WF(k) is the weighting factor for factor k, where $k = 1-15$. The 15th factor is graph scoring, for example. $\text{WF}_i(k)$ is the weighting factor for factor k of word i in a sentence with N words in it. It is that simple: your extractive summary is simply the set of sentences with the highest Total_Sentence_Score until their length reaches your limit of summary length.

2.2.3 Other Considerations in Extractive Summarization

Now that we have seen how to construct an extractive summarization at the word, phrase, and sentence level, we turn our attention to some of the text analysis and natural language processing nuances associated with generating even more accurate summaries. As with any data analytics task, summarization often benefits from careful treatment of the data, which includes data standardization, normalization, and imputation. Standardization for text analytics largely focuses on ensuring that two words which are identical are always equated, and two words which are not the same are never equated. Accents are a strong potential factor on this. For example, consider the word "resume" as in to restart, and compare it to the word "résumé" as in a short biography of a person. What happens if we find the word "résume" or the word "resumé" in our document set? In this case, we might differentiate them based on their part of speech; that is, "resume" should be a verb and "résumé" should be a noun. English as it is practiced in speech, however, causes a wide variety of potential issues here. This is because English, perhaps more than other languages, so readily incorporates neologisms, slang, idiomatic expressions, and part-of-speech migrations. For the latter, think of the terms "ask" and "spend," which are much more commonly used as nouns these days than, say, 20 years ago, particularly in corporate documents. "What's the big ask here?" would have been laughed out of a conversation not so long ago, and if you said, "what's our spend here?" you may have got an open-mouthed stare in the 1990s. We are not the language police, and word use is accounted for by word frequencies, TF*IDF, and other metrics. But that is the point: in English, unlike some other languages, there is no language police, and so the part-of-speech frequencies for specific terms change over time, which certainly makes standardization difficult.

Data normalization, as applied to words, includes stemming, truncation, and lemmatization. The first two are the flip sides of the same coin and are often used interchangeably even though they are functionally different. Stemming is a process of cropping a word down to its "stem"; that is, the part of the word that remains after any prefixes or suffixes (together "affixes") is removed. A stemming algorithm operates on a single word, so it can work on multiple words, phrases, sentences, etc., in parallel since it has no understanding of the context around the words it is working on. Stemmers, therefore, do not have meaning of the part of speech the word is assigned to. Although a hybrid stemmer could be given this information if/as needed, the stemming algorithm may not be able to use this information to its advantage anyway, as it might have no way to distinguish between different parts of speech. Simple stemmers are relatively easy to implement since affixes are well-known. They are, however, lower accuracy than most lemmatizers discussed below. Truncation is the functional use of a stemmed word to collect all related words, typically for searching or other query. For example, suppose the word "diving" was used in a search query. The stemmed version of this is "div-" and thus the untruncated versions are of the form "*div*" where * is an affix. Examples of expanding this truncation are dive, dives, and diving. Note that the word "dove," a conjugation of the verb "to dive," is not possible to derive from the truncation (stem) of "to dive" due to the change of the vowel within the stem.

This is an example of how stemming differs from lemmatization. With lemmatization, at least a modicum of semantics is brought into play. Lemmatization algorithms may use an ontology, a vocabulary/dictionary, and/or morphological analysis of the word not only to remove affixes (which stems the word) but also to provide the base of the word. In the case of "diving," then, the base is "dive" (or the infinitive of the verb, "to dive"), instead of the stem, "div-." Here, the expansion of the lemma into the entire set of derivationally related words include "dive," "dives," "diving," and "dove." Another good example is words that have matching meanings but quite different spellings. For example, the words "best" and "better" both have the word "good" as their lemma. The stem of "best" is simply "best," and the stem of "better" is simply "better," meaning even those two words are not related in stemming. However, the word "sprinting" has the word "sprint" as both its stem and its lemma. Thus, the primary difference between lemmatization and stemming is that lemmatization handles irregular forms of words, while stemming does not.

Irrespective of these differences, lemmatization and stemming can be used to reduce the word set in a document and make the 15 factors described in the previous section even more effective at selecting the optimal set of phrases for extractive summarization. In our experience, the factors word frequency, TF*IDF, lexical similarity, and sentence rank work well for extractive summarization [Ferr13] and perform even better when combined with stemming or lemmatization. However, a bigger advantage for such word normalization may come at the macroscopic level, which, to this point in the chapter, we have not really addressed. One of the problems with the page rank algorithm is that all of the most connected documents will point to each other, and so they will outweigh all of the other documents at the top of the set of search results. This creates a search "echo chamber" of all the same documents being the best search results, irrespective of the search query. A similar problem can arise for the Total_Sentence_Score method (Equation (2.12)) since the phrases with the highest scores may end up biasing the extractive summarization to preferentially select them at the expense of overall summary balance. This is an effect we call a **macro-level summarization failure**, and we provide an example of one next.

Suppose you have a friend who writes a blog posting about their recent travels. Your friend is into visiting National Parks in the US, and on their recent trip, they visited Voyageurs (yes, it is a national park), Rocky Mountain, Arches, Zion, and Grand Canyon National Parks. In their blog, each of these was equally important to your friend and was mentioned no less than 12 times each. Since none of them was mentioned in the title, a lot of the Total_Sentence_Score weighting differences between the names of the five National Parks visited came from the relative frequency of their names in the article compared to the reference documents. In the reference documents, "Voyageurs" occurs less than 10% as often as "Rocky Mountain," "Arches," "Zion," or "Grand Canyon." Thus, the IDF term is at least $10\times$ as high for "Voyageurs" as it is for any of the other parks. As a consequence,

9 out of the 15 sentences selected for the extractive summarization include "Voyageurs" and we get a highly unbalanced summary of the five-park trip.

What do we do to provide a better **macro-level summarization balance**? One approach is to selectively down-weight terms once they have been selected. One intuitive means of doing this is through regularization, wherein the Total_Sentence_Score is adjusted by multiplying a regularization factor, λ, with the sum of occurrences of each of the terms in any sentences that now occur in the summary: this is termed $\text{num}_i(S)$ for each of the N terms in the sentence being regularized, as they occur in the set of sentences already part of the summary. In this way, each sentence score is updated after the summary is appended. The regularization sum penalizes selecting more sentences like the ones already part of the summary, as shown in Equation (2.13):

$$\text{Regularized\_Sentence\_Score} = \sum_{i=1}^{N} \prod_{k=1}^{15} \text{WF}_i(k) - \lambda \sum_{i=1}^{N} \text{num}_i(S).$$

(2.13)

Thus, in Equation (2.13), if the term "Voyageurs" occurs once in the current sentence being evaluated (not already part of the summary), and the same term "Voyageurs" already occurs twice in the set of sentences that are already in the summary, then $\text{num}_i(S) = 2$ for the term on the right-hand part of the equation. If, in another sentence being evaluated, the term "Voyageurs" does not occur, then $\text{num}_i(S) = 0$. For the National Park example given, the regularization term, in fact, eliminates five sentences containing "Voyageurs" from the final summary. The four sentences containing "Voyageurs" are still an appreciable percentage (27%) of the 15 sentences in the final summary but substantially less than the original 60% before regularization.

Other variants of Equation (2.13) are, of course, possible. The most comprehensive approach is simply to retabulate every Total_Sentence_Score for the remaining sentences in the document each time a sentence is selected for the summary. Here, the (now smaller) document is rated as new. This, however, may not be as effective in preventing the type of "rare word lock" we are concerned with for macro-level summarization balance as the Equation (2.13) listed. A second way to modify Equation (2.13) is to multiply the words already in the summary (which are multiplied by the regularization term) by their Total_Sentence_Score. This results in the regularized sentence score for each remaining sentence (not already in the summary) as given by

Equation (2.14):

$$\text{Regularized\_Sentence\_Score} = \sum_{i=1}^{N} \prod_{k=1}^{15} \text{WF}_i(k)$$

$$- \lambda \sum_{i=1}^{N} \prod_{k=1}^{15} \text{WF}_i(k)\text{num}_i(S). \quad (2.14)$$

Another consideration for creating extractive summarizations is how to assess them independently of the means to generate them. For example, using the Total_Sentence_Score (Equation (2.12)) to relatively compare different summarization algorithms is inappropriate if one of the algorithms uses the same equation to generate its summaries − it will tautologically be rated the best summary. We have looked at the means of quantitatively and qualitatively assessing automatic extractive summarization systems [Bati15]. In this work, 22 online available extractive summarization systems were evaluated using the CNN corpus [Lins19a], which is a data set of 3000 news articles whose production was led by our friend and long-time collaborator Professor Rafael Lins of Recife, Brazil, together with his team at the University of Pernambuco. The CNN corpus was created specifically to allow us to perform exact comparison of the extracted phrases against those deemed most relevant to the summary by multiple human experts. The CNN articles chosen had the additional advantage of having summary sentences as part of the article, affording author/editor provided ground truthing. Thus, the CNN corpus evaluation compares the number of phrases selected by the specific summarization system to that of the human-provided "gold standard." This provides both a quantitative (percent matching) and qualitative (humans ground truthed the summaries) approach to rating the summarization systems.

Not content with only a single means of comparison, we also employed the well-known ROUGE (recall-oriented understudy for gisting evaluation) [Lin04] approach to quantitatively assess extraction summarization approaches. ROUGE measures the content similarity between system-developed summaries and the gold standards provided for the test documents. ROUGE-2 was used on the basis of its reported high correlation with the output of expert (human) annotators [Lin04]. In another study [Oliv16], we directly compared 18 sentence scoring techniques, variants on the 15 from earlier in this chapter, to determine the relative importance of individual sentences for extractive single-document and multi-document summarizations. Importantly, in this work, we showed the value of hybridizing (combining) two or more of these sentence scoring techniques, which is work that helped lead to the more general Total_Sentence_Score based technique elaborated here and summarized by Equation (2.12). Combinatorial, also called ensemble or hybrid, techniques were shown in most cases to outperform the results obtained by individual sentence scoring methods, and these techniques were found to compete in summarization quality with state-of-the-art systems such as the 22 evaluated elsewhere [Bati15].

Extending the hybrid approach even further, we decided to deconstruct the manner in which TF*IDF is defined and determined if linear combinations of the different variants of TF*IDF could outperform the individual TF*IDF approaches [Sims17a]. We therein defined a total of 112 TF*IDF equations created by using a combination of 14 IDF equations for each of 8 TF equations. The eight variants of TF were power, mean, normalized log, log, normalized logs, normalized mean, normalized power, and normalized powers of the TF. The 14 variants of document frequency were the normalized log of the sums, the normalized sum of the logs, the sum of powers, the power of sums, mean, normalized sum of logs, normalized log of sums, normalized sum of powers, normalized sums of powers, sum of logs, log of sums, normalized mean, normalized power of sums, and normalized powers of sums [Sims17a]. Our results showed that using weighted combinations of these 112 TF*IDF algorithm provides some improvement. The results also indicated that there are roughly 20 degrees of freedom in these 112 measurements, which hints that it may be better to compute non-linear combinations (multiplications, divisions, powers, etc.) of these TF and IDF calculations for even greater improvement [Sims19].

Further research focused on features and methodology employed for the design and production of the CNN-corpus, mentioned above as a medium-to-large scale corpus for testing single document extractive text summarization [Lins19a]. The documents, as the name indicates, are news articles. Importantly, short, bulleted summaries (a form of minimum gold standard) and article classes are provided. A more than 3000-article English corpus was produced by 2019, and a smaller Spanish corpus was also produced [Lins19b]. The short, bulleted summaries are abstractive summaries, and the extractive summary was produced by consensus of multiple texts in English, and each of them has an abstractive and an extractive summary. The corpus allows quantitative and qualitative assessments of extractive summarization strategies. The corpus was first used as the basis of an international competition for best summarization approach at the ACM Document Engineering Symposium in Berlin, Germany, in September 2019 (the *DocEng'19 Competition on Extractive Text Summarization*) [Lins19c] and is being used in research initiatives for improving extractive summarization techniques and automatically generating abstractive summaries from extractive ones. The CNN corpus, with the original texts, their highlights, gold-standard summaries, and all its annotated versions, are freely available by request for research purposes and will be used in continued competitions at DocEng.

2.2.4 Meta-Algorithmics and Extractive Summarization

As hinted at by the hybrid approaches described above, extractive summarization can benefit significantly from combined, ensemble, or hybrid methods (the terms frequently used interchangeably). In an earlier book covering **meta-algorithmics** (advanced forms of combinatorial algorithms), we introduced some means to use meta-algorithmics to improve the score of a summarizer. In this section, we revisit that experiment [Sims13], accounting for some of the important methods for improving summarization accuracy described above. Perhaps the most important among them is the Sentence_Position_Factor, which tends to favor extracted sentences at the beginning or end of a document.

We begin our synopsis and extension of the previous work [Sims13] with a consideration of training data. The application of meta-algorithmic patterns to extractive summarization relies on the existence of proper ground truthed or labeled training data. Collecting training data, however, is often expensive, time-consuming, and perhaps even contentious (humans do not always agree on their interpretations, as we also found out when compiling the CNN corpus, described above). Additionally, the ground truth collected may have been accumulated with a specific type of intelligent system design envisioned, which means that certain meta-data may have been omitted from consideration. This will almost certainly reduce the value of this training data for related applications, which often are the most valuable over time. In general, then, the rule for ground truthing training data is "if it is reasonable to do so, collect as much contextual data, or meta-data, as possible."

Meta-algorithmics, meaning **advanced combinatorial approaches**, are not only useful for the "forward" problems of machine learning; that is, for improving overall system accuracy, performance, robustness, and/or cost: they can also be used to help optimize the process for collecting data in the first place. We have earlier addressed some of the ways in which we can use re-architecting the process by which ground truth data is collected with the aim of providing a more accurate text summarization system [Sims17b]. In this section, we will address how meta-algorithmics interaction with optimization (to the point of overfitting the training data) depends on how sophisticated the individual summarization engines are.

For combinatorial summarization using two or more summarization engines together, ranking the individual engines in order of confidence is often highly useful. One means of achieving this ranking is to have each summarization engine provide a summary, and then have humans evaluate each summary and rank them relatively. This approach, like most manual ground truthing approaches, is both time-consuming and provides only a binary (paired) comparison. This "binary" approach is unsatisfying, however, as we cannot act on the decision other than to select one summarizer over another summarizer. This has some value for sets of two or three summarizers, but it has diminishing value for more than three summarizers. We suggest, instead, using a meta-algorithmic approach, wherein the summarizers provide the same type of output (in this case, the same number of sentences, $N = 10$) and, thus, can be scaled to any number with linear, rather than geometric, processing requirements.

Summarization can be extractive or abstractive, as noted above. Like above, we have decided to implement the more commonly employed extractive technique, which simply replicates the original text that is calculated to be the most salient for the summary, such as when employing Equation (2.12) above. The extracted sentences, ranked in order of their overall saliency weight, *are* the summarization. Table 2.4 illustrates the ranked order of sentences for three summarizers for a sample document composed of 35 sentences. In this single-document example, Summarizer 1 selected sentence 4 as its most salient sentence, followed by sentence 7, and so on.

The original text (an article) used for this example contains 35 sentences, and the three individual summarizers (1, 2, and 3) employed each selected the 10 most salient sentences and they were assigned weights in inverse order to the ranking (the highest ranked sentence receives weight = 10.0, the second highest ranked sentence receives weight = 9.0, and so forth). Next, human volunteer evaluators are presented with the original text (all

Table 2.4 Sentences selected by three different summarization engines and their relative ranking.

Rank	Summarizer 1	Summarizer 2	Summarizer 3	Weighting
1	4	7	1	10.0
2	7	1	14	9.0
3	3	6	4	8.0
4	14	29	3	7.0
5	25	19	7	6.0
6	9	4	19	5.0
7	1	5	25	4.0
8	33	3	30	3.0
9	19	33	9	2.0
10	35	14	24	1.0

Table 2.5 Sentences in ranked order as output by the human evaluators, and their corresponding relative human weighting (RHW), ranging from 1.0 to 15.0 (weighting reverses the ranking).

Sentence number	Relative human ranking
1	15.0
7	14.0
3	13.0
14	12.0
4	11.0
6	10.0
9	9.0
19	8.0
33	7.0
35	6.0
5	5.0
25	4.0
29	3.0
24	2.0
30	1.0

35 sentences) along with the complete set (the union) of all sentences ranked by all summarizers (this is less than 35 sentences). The volunteers then select what they think are the most relevant 10 sentences and order them from 1 to 10. A score of 1 indicates what to them the most important sentence in the article is, and this receives a relative human ranking of 15.0, as shown in Table 2.5.

In the example, this set of all significant sentences includes only the 15 sentences {1, 3, 4, 5, 6, 7, 9, 14, 19, 24, 25, 29, 30, 33, 35}, which is indeed far less than the 35 total sentences. Thus, 20 of the 25 sentences are not included in any of the three summaries. We can see from this that there is a tendency to select sentences from near the beginning and the ending of the article, in agreement with sentence location weighting discussed earlier in this chapter. Table 2.5 provides the ranked ordering of these 15 sentences as judged by the human evaluators. The highest score was given to sentence 1, the next highest score to sentence 7, and so on. The lowest score (1.0) was given to sentence 30, which was deemed the least significant of any of the 15 selected sentences.

A second, quantitative, means of evaluating the summarizers is also reviewed here. This is illustrated in Table 2.7, where the total weight is computed for each summarizer j, where $j = 1, \ldots, S$, and $S =$ number of summarizers, by simply performing the following operation

Table 2.6 Weight (weighting of rank multiplied by the relative human ranking; thus, a max of 150.0) of each individual sentence selected by each of the three summarizers, and total weight of each summarizer (sum of all weights). The overall weight of Summarizer 3, at 600.0, is slightly higher than that of Summarizer 1, at 596.0. The lowest weight is for Summarizer 2, at 564.0.

Rank of the sentence	(Weighting of rank) × (relative human ranking)		
	Summarizer 1	**Summarizer 2**	**Summarizer 3**
1	110.0	140.0	150.0
2	126.0	135.0	108.0
3	104.0	80.0	88.0
4	84.0	21.0	91.0
5	24.0	48.0	84.0
6	45.0	55.0	40.0
7	60.0	20.0	16.0
8	21.0	39.0	3.0
9	16.0	14.0	18.0
10	6.0	12.0	2.0
Total weight	**596.0**	**564.0**	**600.0**

(Equation (2.15)):

$$\text{TW}_j = \sum_{i=1}^{N_S} W(i) \times \text{RHW}(S(i,j)). \qquad (2.15)$$

Here,

TW_j = total weight for summarizer j;

N_S is the number of sentences in each summary;

$W(i)$ is the weight associated with rank i (in our example, this is simply the quantity $N_S + 1 - i$, as shown in the "Weighting" column of Table 2.4);

$S(i,j)$ is the sentence number associated with rank i for Summarizer j (e.g., $S(3,5) = 7$ and $S(8,1) = 33$ in Table 2.4);

$\text{RHW}(S(i,j))$ is the RHW of the sentence identified by $S(i,j)$.

For example, $\text{RHW}(S(3,5)) = 14.0$ and $\text{RHW}(S(8,1)) = 7.0$ as shown in Table 2.5, in which the left column are the values of $S(i,j)$ and the right column are the values of $\text{RHW}(S(I,j))$. Table 2.6 is therefore populated with the products of the relevant RHW and W rankings. The sum shows that the overall weight of Summarizer 3 – that is, 600.0 – is slightly greater than that of Summarizer 1, which is 596.0. The lowest weight is for Summarizer 2, at 564.0. This indicates that, for this document, Summarizer 3 provides the best overall results of the three, but Summarizer 1 is not much different. Thus, if we followed the meta-algorithmic pattern of constrained substitution [Sims13], we could certainly use Summarizer 1 in place of Summarizer 3 if we wished to do so for other reasons than accuracy of summarization (e.g., cost, performance, availability, existence of a licensing agreement, etc.).

The RHW approach outlined here offers a second level of comparison among summarization algorithms. It also ensures a blind evaluation since the person providing the sentence order does not know which sentences have been selected by the summarizers. The RHW approach provides quantitative comparative data. In the simple example shown above, Summarizers 1 and 3 are shown to be very similar in overall weighting and, relatively speaking, more heavily weighted than Summarizer 2 when used in combination. For a much larger sample set, such relative differences would be very important – if consistent with the results of this single example document, they would indicate that Summarizers 1 and 3 are more or less interchangeable in quality and value. These differences would also indicate that Summarizer 2 should not be used in place of Summarizers 1 and 3.

Additional value provided by the RHW approach is that, in providing a full ranking to all of the sentences in all of the combined set of summarizations, the RHW method allows us to explore many different combinations of two or more summarizers (that is, meta-algorithmic and other ensemble patterns). One of the simplest meta-algorithmic patterns is the voting pattern, which is directly supported when employing the RHW approach. This pattern, when applied to summarization, consists of adding the relative weighting for the ranking of each individual summarizer for each sentence. These values are tabulated in the second column of Table 2.7. To illustrate how this proceeds, consider sentence 7 in the original article. For Summarizer 1, sentence 7 is ranked 2nd (9.0 weighting); for Summarizer 2, it is ranked 1st (10.0 weighting); and for Summarizer 3, it is ranked 5th (6.0 weighting). The combined weighting, 9.0 + 10.0 + 6.0, is 25.0 and is the highest of any sentence. Similarly, sentence 1 (23.0) and sentence 4 (23.0, with the tie-breaker being the second ranking value, which is higher for sentence 1) are the next two highest weighted sentences by the combination of summarizers. This {7, 1, 4} is different from the ranking provided by the human evaluators; namely, {1, 7, 3} (Table 2.5). If the ranked order of a given summarizer were the same as the ranked order of the human evaluators, the maximum total weight, TW_{max}, is obtained. This weight is determined by Equation (2.16).

$$TW_{max} = \sum_{i=1}^{N_S} W(i) \times RHW(i). \qquad (2.16)$$

For the given example, TW_{max} = 660.0. As shown in Table 2.7, the (equal) voting combination of Summarizers 1–3 results in a much improved summarizer, for which the total weight is 627.0. This is 45% closer to the

Table 2.7 Weight (weighting of rank multiplied by the relative human ranking) of each individual sentence selected by the combination of Summarizers 1, 2, and 3, and the total weight of the summarizer (sum of all weights). The combined summarizer substantially outperforms any of the individual summarizers, with a total weight of 627.0 (out of a maximum possible 660.0), compared to 600.0 or less for each of the individual summarizers.

Rank	Sentence (sum of ranks)	(Weighting of rank) × (relative human ranking) of the non-weighted combination of Summarizers 1−3
1	7 (25.0)	140.0
2	1 (23.0)	135.0
3	4 (23.0)	88.0
4	3 (18.0)	91.0
5	14 (17.0)	72.0
6	19 (13.0)	40.0
7	25 (10.0)	16.0
8	6 (8.0)	30.0
9	29 (7.0)	6.0
10	9 (7.0)	9.0
Total weight		**627.0**

ideal score of 660.0 than the best individual summarizer – that is, Summarizer 3 with a score of 600.0.

Importantly, other meta-algorithmic/ensemble patterns can also be readily applied to the relatively ranked human evaluation data. For example, the meta-algorithmic weighted voting pattern uses a weighted combination of Summarizers 1−3. The weights for the individual summarizers in the combination can be determined as proportional to the inverse of the error, e, in which case the weight of the jth summarizer, W_j, is determined from Equation (2.17):

$$W_j = \frac{\frac{1}{e_j}}{\sum\limits_{i=1}^{N_{\text{SUMM}}} \frac{1}{e_i}} \tag{2.17}$$

where N_{SUMM} is the number of summarizers and error is defined as given in Equation (2.18):

$$e_i = \text{TW}_{\max} - \text{TW}_i. \tag{2.18}$$

For the specific problem at hand, error $e = 660.0 - \text{TW}$.

For simplicity here, let us assume that the error, e, of each of the three summarizers on the training data is the same as we observed in this example. Then, using the equation above, the weighting of the three summarizers are $\{0.366, 0.244, 0.390\}$ for Summarizers $\{1, 2, 3\}$. The effect of weighting the combination of the summarizers is described by Equation (2.19):

$$\text{SumOfRanks}_i = \sum_{j=1}^{N_{\text{SUMM}}} W_j \times W(i). \tag{2.19}$$

These values are shown in the parentheses in the second column of Table 2.8. This weighted voting approach results in a total weight of 628.0, which is 46.7% closer to the best possible score of 660.0 than the best of the individual summarizers.

In this example, the meta-algorithmic voting approaches were shown to improve the agreement between the automated extractive summarization and that provided by human ground truthing. Moreover, the meta-algorithmic approach was shown to be consistent with a different type of ground truthing, in which the human expert simply ranks the sentences in order of importance according to their (expert) opinion. Here, the appropriate ground truth is also extractive, meaning that the sentences are ranked for saliency. This is advantageous to the normal ranking of individual summarizations approach because (1) it is scalable to any number of summarizers, (2) it is innately performed with the human expert blind to the output of the individual summarizers, and (3) it supports the advantageous voting-based meta-algorithmic patterns illustrated herein.

We now repeat this same summarization experiment, using the same example document, when we employ Equation (2.12), the Total_Sentence_

Table 2.8 Weight (weighting of rank multiplied by the relative human ranking) of each individual sentence selected by the weighted combination of Summarizers 1, 2, and 3, and the total weight of the summarizer (sum of all weights). This combined summarizer also substantially outperformed any of the individual summarizers, with total weight of 628.0, compared to 600.0 or less for each of the individual summarizers. The results here are not a significant change from those of Table 2.7.

Rank	Sentence (sum of ranks)	(Weighting of rank) × (relative human ranking) of the weighted combination of Summarizers 1−3
1	7 (8.074)	140.0
2	4 (8.000)	99.0
3	1 (7.560)	120.0
4	3 (6.390)	91.0
5	14 (6.316)	72.0
6	19 (4.146)	40.0
7	25 (3.756)	16.0
8	9 (2.610)	27.0
9	6 (1.952)	20.0
10	29 (1.708)	3.0
Total weight		**628.0**

Score, as an addendum to each of the three summarizers. The Total_Sentence_Score is strongly influenced by the Sentence_Position_ Factor, described by Equation (2.8). With this accounted for, we get the following table (Table 2.9), which changes some of the sentence ranks from Table 2.4 above.

Noticeably, sentence 1 has moved up for Summarizer 1 and remains the top choice of Summarizers 2 and 3. Next, the relative human ranking (which is, of course, unchanged from Table 2.5) is used together with the summarizer output of Table 2.9 to generate the RHW-based information in Table 2.10. The summarizers have weights of {606.0, 557.0, 600.0}, which is not a large change (<1% overall) from those of Table 2.6, or {596.0, 564.0, 600.0}.

Next, Equation (2.15) is used to generate the summed ranks of the sentences, which have now been, at least in some cases, reordered by the various considerations of Equation (2.12). These new rankings are collected in Table 2.11.

The results in Table 2.11 are substantially better than those for any individual summarizer. 635.0 is a 54% reduction in error (that is, difference from 660.0) in comparison to the best individual summarizer, Summarizer 1. The top five sentences, while slightly mixed in order, are the same as the top five sentences provided by the human experts.

There is not much room for improvement. In fact, when we further modify the summarizers by using weighted voting, following Equation (2.17), we do not get a further improvement in the summarizer's accuracy (that is, agreement with ground truth), as shown in Table 2.12. The normalized weights of the summarizers, calculated as being proportional to the inverse of their error rates, are {0.412, 0.216, 0.371} for summarizers {1, 2, 3}. With this weighting, the total weight of the summarization actually drops from 635.0 to 626.0. This value still reduces the error by 37.0%

Table 2.9 Sentences selected by three different summarization engines and their relative ranking, where the Sentence_Position_Factor is implemented as per Equation (2.8).

Rank	Summarizer 1	Summarizer 2	Summarizer 3	Weighting
1	4	1	1	10.0
2	3	7	14	9.0
3	7	6	4	8.0
4	1	29	3	7.0
5	35	19	7	6.0
6	14	33	19	5.0
7	33	4	25	4.0
8	25	5	30	3.0
9	9	3	9	2.0
10	19	14	24	1.0

Table 2.10   Weight (weighting of rank multiplied by the relative human ranking) of each individual sentence selected by each summarizer, and total weight of the summarizer (sum of all weights). The overall weight of Summarizer 1, at 606.0, is for this situation slightly higher than that of Summarizer 3, at 600.0. The lowest weight is for Summarizer 2, at 557.0. Compare to Table 2.6.

Rank of the sentence	(Weighting of rank) × (relative human ranking)		
	Summarizer 1	Summarizer 2	Summarizer 3
1	110.0	150.0	150.0
2	117.0	126.0	108.0
3	112.0	80.0	88.0
4	105.0	21.0	91.0
5	36.0	48.0	84.0
6	60.0	35.0	40.0
7	28.0	44.0	16.0
8	12.0	15.0	3.0
9	18.0	26.0	18.0
10	8.0	12.0	2.0
Total weight	**606.0**	**557.0**	**600.0**

Table 2.11   Weight (weighting of rank multiplied by the relative human ranking) of each individual sentence selected by the combination of Summarizers 1, 2, and 3, and the total weight of the summarizer (sum of all weights) for Tables 2.9 and 2.10. The combined summarizer substantially outperformed any of the individual summarizers, with a total weight of 635.0, compared to 606.0 or less for each of the individual summarizers.

Rank	Sentence (sum of ranks)	(Weighting of rank) × (relative human ranking) of the non-weighted combination of Summarizers 1−3
1	1 (27.0)	150.0
2	7 (23.0)	126.0
3	4 (22.0)	88.0
4	3 (18.0)	91.0
5	14 (15.0)	72.0
6	19 (12.0)	40.0
7	33 (9.0)	28.0
8	6 (8.0)	30.0
9	29 (7.0)	6.0
10	25 (7.0)	4.0
Total weight		**635.0**

compared to that of the best individual summarizer but does not improve the accuracy over simple voting.

The results of Tables 2.11 and 2.12, taken together, indicate that weighted voting provides no advantage over simple voting in this case. This is, almost assuredly, due to the fact that adding complexity to the summarizers (using the Total_Sentence_Score) turned each of those summarizers into

Table 2.12 Weight (weighting of rank multiplied by the relative human ranking) of each individual sentence selected by the weighted combination of Summarizers 1, 2, and 3, and the total weight of the summarizers (sum of all three weights). The weights of the summarizers, from Equation (2.17), are $\{0.412, 0.216, 0.371\}$. This combined summarizer also substantially outperformed any of the individual summarizers, with total weight of 626.0, compared to 606.0 or less for each of the individual summarizers. However, it did not improve upon the results on Table 2.12, giving an example of how meta-algorithmics may not benefit already meta-algorithmic summarization engines used in combination. See text for details.

Rank	Sentence (sum of ranks)	(Weighting of rank) $\times$ (relative human ranking) of the weighted combination of Summarizers 1−3
1	1 (8.754)	150.0
2	4 (7.952)	99.0
3	7 (7.466)	112.0
4	3 (6.737)	91.0
5	14 (5.615)	72.0
6	19 (3.563)	40.0
7	33 (2.744)	28.0
8	25 (2.720)	12.0
9	35 (2.472)	12.0
10	6 (1.728)	10.0
Total weight		**626.0**

a meta-algorithmic summarizer. Thus, the additional advantage of adding complexity to the meta-algorithmic algorithm was already incorporated into the summarizers themselves.

Next, we repeated the experiment for a set of 20 documents roughly the size of the original document (that is, 20−40 sentences). The results for all approaches are tabulated in Table 2.13. Taking Table 2.13 as a whole, adding in the Total_Sentence_Score reduced the error percentage by a mean of 1.3%, from 9.3% to 8.0%, which is a 14.3% reduction in the actual error rate. The individual summarizers improved by a mean of 1.3%. The voting approach improved by 2.0% when Total_Sentence_Score was incorporated into the summarizers, reducing the error percentage from 6.2% to 4.2%. The weighted voting approach reduced the error percentage from 4.7% to 3.8%. Thus, using sentence scoring in addition to weighted voting reduced the error rate from 8.8% to 3.8%, in the mean (Table 2.13).

This section describes a wide variety of processes by which to perform extractive summarization. We focused on extractive summarization because it offers us functional advantages over abstractive summarization, including the following:

(1) It preserves the words of the author; therefore, it contains quotable material.

Table 2.13 Results for 20 documents, with 20−40 sentences each, following the processes for Tables 2.4−2.12. Adding in the Total_Sentence_Score reduced the error percentage to a mean of 1.2%, from 9.3% to 8.0%, or a 14.3% reduction in the actual error rate.

Approach	Total weight mean (μ)	Total weight STD (σ)
Original summarizers 1, 2, and 3		
Summarizer 1	590	18
Summarizer 2	552	22
Summarizer 3	602	17
Voting, $\{1,2,3\}$	619	15
Weighted voting, $\{1,2,3\}$	629	18
Original summarizers + Total_Sentence_Score (Equation (2.12))		
Summarizer 1	602	20
Summarizer 2	559	13
Summarizer 3	608	22
Voting, $\{1,2,3\}$	632	20
Weighted voting, $\{1,2,3\}$	635	17

(2) It preserves the linguistic style and preferences of the author; therefore, readers of the summary will still be able to get a sense for how the author wrote.

(3) It is in continuum with other levels of repurposing text, including keyword generation, indexing, abstracting, and other functional analytics.

(4) The types of errors made are generally corrected by lengthening the summarization, whereas errors in abstractive summarization may not be corrected with lengthening.

(5) If specific terms, phrases, sentences, and other passages (e.g., specific quotes) are required to be kept in the summary, these are already extractive in nature.

In this section, we described the means to relative weight different terms, phrases, sentences, and passages within a document for salience to the extractive summary. These 15 weighting approaches provide the text analyst tools to evaluate many summarization approaches for the task at hand. In the next section, we consider some of the machine learning aspects of summarization, which bridges us over to abstractive summarization.

2.3 Machine Learning Aspects

One of the primary areas for machine learning in summarization is for forming abstractive summaries. This is a huge topic, which will only be addressed at a superficial level here, since functional summarization, as noted above, usually requires extractive summarization. Machine learning will continue to be especially important for interpreting idiomatic expressions, slang, jargon, and other specialized usage for both extractive and abstractive summarization. A good example of the value of machine learning will be in learning the reasons for particular word usage, which may not be straightforward to do with extractive methods such as those outlined in this chapter.

One example of where machine learning will be helpful is the case in which there are overloaded terms which cause asymmetric changes in language usage. For example, the word "caustic" is used in place of the word "basic" frequently in documents concerned with chemistry. Technically, the word caustic means "capable of burning or corroding tissue by chemical activity," which means it could be either "basic" or "acidic," since either of these two types of compounds are caustic. However, the term "basic" is heavily overloaded since it is one of the more, pardon the word play, basic words in the English language. Acidic, quite frankly, is not. From a usage standpoint, "basic" can be represented by the word "alkaline," which is not an overloaded operator (unless you count Al Kaline, the former Tiger great who passed while this chapter was being written, but that seems a bit of a reach!) and is therefore more precise. However, the fact of the matter is that "basic" and "acidic" are quite different in terms of their overall usage: "basic" is generally used for a different meaning than "alkaline," while "acidic" is generally used to imply a low-pH caustic substance (there can be an acidic personality, of course).

In order to use machine learning in summarization, generally, very large data sets are required. Why is this? It is because, depending on your reference source, and whether or not you include obsolete words, English as a language contains somewhere between 150,000 and 250,000 words. At the higher end, that is twice the number of words as are in this book. As we know, though, these words are not randomly distributed, and so you may, in the mean, only go 50 or 100 words before encountering your next instance of "the," but you may go the rest of your lifetime before you encounter the next instance of "nudiustertian," "kakorrhaphiophobia," or "antidisestablishmentarianism." Words are not equally used (for example, they may follow Zipf's Law, or perhaps more historically accurate, Estoup's Law), and even if so they are not equally distributed. The frequency of many words in the English language (in fact, the vast majority of them) in normal text is less than one in a million words. In order to generate a meaningful estimate for such important analytics as co-occurrence and correlation of words, therefore, we need training sets that are very large. Tens of millions of words are needed to get even cursory estimates for relevant statistics about words outside of the list of 5000 or 10,000 most commonly used words. Fortunately, it is increasingly easy to collect, organize, and access very large data sets.

In order for machine learning approaches to be properly assessed, they must be compared to their variants and to alternate machine learning approaches with the same ground truth data used for training each individual approach. Variants include different macro-algorithmic approaches, which means largely different families of algorithms. Neural networks, evolutionary algorithms, immunologically inspired algorithms, SVMs, Bayesian methods, and ensemble methods are examples of families of analytic algorithms. Among each of these families, there are typically several variants of interest. Deep learning approaches [LCu15] have garnered a lot of interest, and a lot of publications, in the past few years; rightly so. Deep learning has benefitted from continued increases in processing and storage, allowing neural networks to add considerably more input, hidden, and output nodes [Schm15]. This has enabled neural networks to incorporate advanced methods for massively increasing the number of features generated for use in classification, recognition, and other advanced deep learning in imaging [Kriz12]. As applied to text, neural networks and deep learning can be used to relate sequences of words, represented as adjacent input nodes, to specific meaning, represented as different output nodes. Another important family of machine learning approaches with a number of variants is the ensemble methods. Ensemble methods invoke two or more classification or recognition approaches simultaneously and intelligently combine them to provide higher value (usually higher accuracy, but it could also be higher robustness, higher performance, etc.) output than that provided by any of the individual approaches. Traditional ensemble approaches involve relatively simple voting or weighted voting techniques and include stacking, bagging, and boosting. Advancements in boosting include the adaboost approach [Freu95] and weighted voting and weighted confusion matrix techniques [Sims13]. These techniques incorporate multiple algorithms, with the overall contribution of each algorithm to the final decision being weighted by the relative confidence in each algorithm. Generally, confidence in an algorithm corresponds to its precision; that is, its expected accuracy for the decision it reports.

Our preference for machine learning design for text analytics, perhaps not surprisingly, is the use of meta-analytic approaches [Sims19]. These approaches combine the high trainability of deep learning algorithms with the central limit theorem and domain coverage advantages of ensemble methods. A typical meta-analytic pattern might use a Bayesian network, a neural network, and a traditional natural language processing (NLP) approach in combination. These methods might provide excellent agreement where they overlap in their training and differential accuracy where their confidence is higher than the other two algorithms. Selecting the highest precision method among several methods typically results in a 15%−50% reduction in error rate for the overall system [Sims19].

2.4 Design/System Considerations

We have introduced four important system-level considerations for summarization in previous sections: (1) scale; (2) structure; (3) time; and (4) macro-level balance. Scale is concerned with the length of text that should be considered as the "atomic" unit for text analytics. Structure is concerned with the process by which we search for relevance to the search. Time is important for acknowledging that the relevance of specific text changes over time. Macro-level balance is concerned with how the summarization is dynamically extracted from its original "reservoir" of possible text. We consider each of these factors in this section and provide some guides for how to take them into account in building a robust summarization system.

Camus's *The Stranger* begins with the famous words "Maman died today. Or yesterday maybe, I don't know." The frequencies of each of these nine words in the novel are $\{35, 11, 9, 60, 5, 13, 1306, 32, 60\}$, which as fractions of the 36,830 words in the novel are $\{0.000950, 0.000299, 0.000244, 0.001629, 0.000136, 0.000353, 0.35460, 0.000869, 0.001629\}$. The surprising one might be "I," which occurs less than only the word "the" in frequency in the novella. Regardless, these data illustrate the uneven coverage of the English vocabulary in the novella and the extent of coverage of the words in the novel by the quote. Summing up the nine words, they represent 1531 words out of 36,830 in the novel. This is 4.16% coverage of the novel or 0.46% per word. This, in turn, means the words cover the novel-equivalent representative terms in just 216.5 words. We term these **expression coverage** and **mean expression coverage**, respectively, and these represent another perspective on the **scale** of a phrase or expression. In general, the lower the expression coverage, and, in particular, the mean expression coverage, the more likely the expression is to contain rare terms in the document of interest. However, this treats the expression, and the document, like a bag of words. It is more likely that we are concerned with the less frequently occurring terms in the particular expression: these are "died," "today," and "yesterday." From those, we get a real sense of the co-occurrence of differentiating terms. The only times in the novel when these three terms are in close proximity are the two times where the protagonist, Mersault, describes his almost insouciant lack of precise information about the timing of his mother's death. At the scale of looking for multiple (in this case, three) relatively infrequent terms that co-occur, this example might tie better to searching than to summarization; however, it does show how sections of text may be pre-organized for query. Irrespective of whether this expression may ever be chosen for a generic extractive summarization, it shows that scale for text analysis generally involves more than single words. This "middle ground" of co-occurrence may thus represent a suitable balance between word-level analysis (exemplified by Equation (2.12)) and semantic analysis, which is simulated in part by co-occurrences.

Structural harmony between the approach for summarization and the goal(s) of summarization is important. With the example from Shakespeare's *Hamlet* earlier, the phrase "To be, or not to be, that is the question," cannot be included in a summary solely based – or, perhaps, even highly based – on sentence scoring techniques such as those accumulated in Equation (2.12). Here, instead of ranking expressions by the relative rareness of their collective words, we are actually looking for external correlation of the properly **scaled** expressions. Thus, we are looking for the relative frequency of an expression across a wide diversity of other documents, such as critiques of *Hamlet*, use of expressions in *Hamlet* in popular culture, etc. As a further example, when Polonius pontificates, "Brevity is the soul of wit," in addition to the irony, he may actually be alluding to summarization. Whether he anticipated this chapter or not, however, he is correct – a summarization that conveys as much understanding about its referent document in less words is generally more highly valued. But, how do we ensure that we have found the right set of expressions in *Hamlet* to convey its plot, its literary influence, and, in this case, its most important set of phrases and sayings. "Alas, Poor Yorick," we must simultaneously consider each of these, along with other concerns like the macro-level balance discussed above and below in this chapter.

Fortunately, for extracting key expressions out of a document, we can provide a ready set of references to find out the best matching expressions between the documents and an appropriate reference set. One such set is a set of critical documents referring to the source document. For example, *Hamlet* is one document, and 10 critical works of *Hamlet* could be compared with *Hamlet* to see which expressions are in common. The expressions in common thereafter make good candidate summarization elements. We performed a throw-away experiment on 100 shorter source documents in Table 2.14, wherein the source document is mined for the 3, 4, 5, 6, 7, or 8 least frequently occurring terms in each sentence of 15 words or more in length, in sequence. For sentences less than 15 words in length, two or more sequential sentences are combined until a minimum of 15 words are reached. Then, the co-occurrences of these $3-8$ words with like-derived expressions in the critical works can be found. After that, exact quotation matches can readily be mined (the sentence in the source document needs to be matched

Table 2.14 Results for a set of 100 documents, together with 1000 critical works about them (10/document), where sentences of 15 words or more, or paired sentences when less than 15 words, are mined for their words (3, 4, 5, 6, 7, or 8) with the lowest frequency in the document. Lower numbers of words (e.g., 3 and 4) are likely to end up with many false positive matches, while larger numbers of words (e.g., 7 and 8) are likely to include many more non-specific (that is, frequently occurring) words. The highest percent quote matching in this simple example is when choosing the five least frequently occurring words in the expression (1−3 sentences). Please see text for details.

# Rare terms co-occurring	# Referent matches	# Quote matches	% Quote matches
3	88,568	5177	0.0585
4	57,342	4322	0.0754
5	40,592	3512	0.0865
6	29,817	1685	0.0565
7	19,953	834	0.0418
8	11,103	457	0.0412

in the critical work to count as a "match" in Table 2.14). The co-occurrence approach has the advantage of providing a fast "search" for matching the two documents (source and critical work). Table 2.14 provides the results for these 100 source documents along with 10 critical works referring to each document (1000 critical works total). The source document sentences (or sequential sentences treated as a single sentence, when less than 15 words in length) are mined for their words (3, 4, 5, 6, 7, or 8) with the lowest frequency in the document. The example given above for *The Stranger* would result in the three words "died," "today," and "yesterday" (never mind that sentence was less than 15 words in length for now).

As the results in Table 2.14 show, using the set of five rarest words to represent the sentence does the most precise job of aligning with actual quotes in the critical works when co-occurrence is used for referent match (column 2 of Table 2.14). Actual quote matches for the five rarest words are 8.65%, with the next highest being for the four rarest words (7.54%). This means that 8.65% of the time there was a match of those five co-occurring words, a direct quote was found. Lower numbers of words (e.g., 3) are likely to end up with many false positive matches, while larger numbers of words (e.g., 7 and 8) are likely to include many more non-specific (that is, frequently occurring) words. The highest percent quote matching in this simple example is when choosing the five least frequently occurring words in the expression (1−3 sentences). There are many matches that are partial quotes or simply shared idiomatic expressions; for example, several of the eight co-occurrence matches were sentences containing the expression "on the other hand," "let the cat out of the bag," or "gave it a run for the money." Shared common multi-word expressions are among the false positives seen.

The results of Table 2.14 also address the **temporal nature of summarization**. The referent documents are critical in determining the match between relevance and summarization. Critical works, such as those addressed in Table 2.14, are likely to quote from the source material and, thus, select the most familiar expressions – like "to be or not to be," "neither a borrower nor a lender be," and "to thine own self be true" – if we do not also impose some form of total sentence weighting as discussed above. If we do entirely separate the summarization from referent documents and use sentence weighting, it is understandable how less important sentences such as "As hardy as the Nemean lion's nerve" and "Gonzago is the duke's name; his wife, Baptista" would be selected for the summary.

This tradeoff between selecting familiar quotes versus selecting expressions containing rare terms (which often specifically identify key people, topics, locations, etc.) is a classic concern of the **macro-level balance** we strive to have in summarization. Not surprisingly, one way to address this issue is to have a certain percentage of the summarization driven by sentence weighting-based methods, with or without regularization approaches (Equations (2.12), (2.13), and (2.14)), and the rest of the summarization driven by quote-matching techniques like those illustrated in this section. The percentage due to one approach versus the other can be optimized functionally; that is, through validation sets.

What we use for validation, however, is also an important consideration in summarization. In order to design an experiment with the proper validation set, we have to carefully consider what the summary is to be used for. Most of this chapter was concerned with selecting the expressions (usually sentences) out of a document that best provide key phrases as part of the summary. The extreme of this approach is of course the TF*IDF approach. However, the query-based approach, which is innately similar to the "familiar quotation" based approach overviewed in this section, can provide summaries that have little to none of the rarest terms in the document. This might be poor for keyword generation but excellent for providing familiarity with the document's public understanding. Still other forms of summarization are feasible. Differential summaries, for example, might be useful for highlighting what has been added to a document over time. Differential summaries can also be used as part of a plagiarism detection process.

Ultimately, the value of a summary is in its usefulness to those choosing to read it in place of the original. In our massively online world, the relative value of a summary may be ascertained from such mundane measurements as clicks or downloads. For a web-based publisher, the amounts of click-throughs and advertising revenue that a summary generates will be an important return on investment (ROI) metric. For users wishing to learn material, the ROI may be based on how well a summary helps them study for and perform well on a subsequent examination. Their ROI may also be how well they can recall the material at a later time. These functional measurements will thus vary from person to person, which suggests that multiple summaries are valuable for any document that may be used for more than one purpose. For example, differential summaries of multiple documents can be used to suggest proper sequencing for the "best next document" to enhance learning. Summarization is, therefore, a key part of an overall approach to organizing, managing, and sequencing document use.

2.5 Applications/Examples

As illustrated throughout this chapter, summarization is used for far more than simply providing a concise representation of a document. Summarization can be used to provide familiar expressions from a document. Summarization can be used to mine indexing, or "keyword," terms from a document. Summarization can be used to provide enhanced learning. We have discussed all of these at one point in the chapter already. However, one additional application for summarization has not yet been covered. This is the use of a summary as a **functional equivalent** of a document.

By functional equivalent, we mean a document that, substituting for the original document, will behave the same as the original. Functionally equivalent summarized documents will appear in the same order in response to the same search queries as the original document. Functionally equivalent summarized documents will provide the same set of indexing words, the same TF*IDF behavior and the same literary "feel" as the original. The indexing words can be preserved through sentence scoring approaches since scoring approaches tend to favor the words that stand out in the original document. However, these same approaches tend to increase the TF*IDF values of these indexing terms since they are preferentially present in the extracted expressions. The macroscopic balancing techniques highlighted by Equations (2.13) and (2.14) can help to offset some of this enhanced TF*IDF behavior; however, it is very difficult to produce a summary that both provides the correct indexing terms and does not change the TF*IDF behavior. As with many other aspects of summarization, there must be a balance stricken here. Fortunately, the third of these three goals – maintaining the same literary "feel" as the original document – can be addressed through the choice of extractive, rather than abstractive, summarization. Thus, we see that the choice of summarization technique can be tied to the downstream use of the summary. The wide range of summarization approaches provides us with this flexibility.

2.6 Test and Configuration

Functional summarization offers significant advantages for test and configuration. Entering into the summarization process with an aim to provide a specific goal or set of goals provides us with the means for testing at the same time. For example, suppose we wish summaries to be used for education, and along with the summarization, we need to provide document sequencing (this is always the case for educational purposes; it is why every class has a syllabus). Whatever algorithm we are using for ordering the original documents should also order the summaries in the same order.

An example of this functional means of assessing summarizers is illustrated in the data of Table 2.15. Ten documents, labeled "A" through "J" alphabetically, are originally ordered from "A" to "J" for reading sequencing as part of a learning plan. Thus, document "A" might provide the most rudimentary material, and document "J" the most advanced material. We have five different summarizers, labeled "1" to "5," which are used to provide extractive summarizations of 20% each of the 10 documents. We wish to evaluate these summarizers for their value without having to use ground truthing. Thus, our functional measurement of summarizer value is to use the document sequencing algorithm to sequence the summarized documents and then compare the document sequencing differences. The difference is assessed by summing up the ordering placement differences. For example, if document "X" is originally the seventh document in sequence, but a summarizer places it as the ninth document in sequence, this is a distance of "2." Summing up over the set of 10 documents, we get a summed ordering distance between any of the summarizers and the original, non-summarized documents. In Table 2.15, the sum of place distances is minimal for Summarizer 4, where "E" is one place different, "F" is one place different, "D" is two places different, "I" is one place different, and "J" is one place different from the original document ordered for the non-summarized documents. This is the minimum summed ordering distance (less than the 8, 10, 14, and 22 of the other four summarizers), thus making Summarizer 4 the "best" summarizer among the five based on the functional measurement of matching suggesting document reading sequence.

Configuring a summarizer, therefore, highly depends on what the summarization will be used for. If the summarization will be used mainly to convey the gist of the document to a human reader, then the sentence weighting approaches culminating in Equations (2.12), (2.13), and (2.14) are generally sufficient. If specific quotes and other memorable text from the document need to be represented in the summary, then the co-occurrence approaches of this chapter should be used in place of, or in addition to, the sentence weighting approaches. Finally, functional means of validating

Table 2.15 Document ordering and ordering distance from the ordering of the original documents for a variety of summarization algorithms. The distance is the place difference between the order of the document for the summarizer for each of the 10 documents. This sum of place distances is minimal for Summarizer 4, where "E" is one place different, "F" is one place different, "D" is two places different, "I" is one place different, and "J" is one place different from the original document ordered for the non-summarized documents. This is the "best" summarizer among the five for retaining document sequencing for learning. Please see text for details.

Document set	Document order	Distance to original	Ranked order
Original	ABCDEFGHIJ	0	N/A
Summarizer 1	ABEDCGHFIJ	8	2
Summarizer 2	BACDFEHJIG	10	3
Summarizer 3	CDABFGHEIJ	14	4
Summarizer 4	ABCEFDGHJI	6	1
Summarizer 5	DEAGCFBJHI	22	5

the specifics of the summarizers, e.g., the relative contribution of the different summarization approaches in a combined ("hybrid") summarization approach, can be employed to avoid the high cost and inconvenience of ground truthing the summarization.

2.7 Summary

To summarize this section, yes, this is in fact a summary of a chapter on how to generate summaries. This chapter describes a wide variety of processes by which to perform summarization, with the primary focus being on extractive summarization versus abstractive summarization. As mentioned above, the reasons for this preference are many, but our primary justifications for favoring extractive summarization are the following:

(1) It preserves the words of the author, and, thus, it is generally capable of providing directly quotable material;
(2) It preserves the linguistic style and preferences of the author, allowing a reader of the summary to still have a sense of how the author writes;
(3) It is better in continuum with other levels of repurposing, including keyword generation, indexing, abstracting, and other functional analytics;
(4) The types of errors made are generally corrected by allowing lengthening of the summarization, whereas errors in abstractive summarization may not be corrected with lengthening;
(5) If specific terms, phrases, sentences, and other passages are required to be kept in the summary, these constraints are already extractive in nature.

Regardless of the choice for summary type, this chapter focuses on the means to relatively weight different terms, phrases, sentences, and passages within a document for salience to the summary. Starting with relative TF, continuing with functional evaluation of summarization approaches, and ending with machine learning approaches, a wide set of candidate summarization approaches are provided. These, in combination, allow the text analyst the opportunity to evaluate multiple summarization approaches concurrently and intelligently select the one that best meets the needs of their application. Why would you not want to have as many tools at your avail as possible?

References

[Basi16]   Basiron H, Jaya Kumar Y, Ong SG, Ngo HC, Suppiah PC, "A review on automatic text summarization approaches," *Journal of Computer Science*, 12, 178-190, 2016.

[Bati15]   Batista J, Ferreira R, Tomaz H, Ferreira R, Dueire Lins R, Simske S, Silva G, Riss M, "A quantitative and qualitative assessment of automatic text summarization systems," *Proceedings of the 2015 ACM Symposium on Document Engineering*, 65-68, 2015.

[Brin98]   Brin S, Page L, "The anatomy of a large-scale hypertextual web search engine," *Computer Networks*, 30, 107-117, 1998.

[Edm61]   Edmundson HP, Wyllys RE, "Automatic abstracting and indexing—survey and recommendations," Communications of the ACM 4.5 (1961): 226-234.

[Edm69]   Edmundson HP, "New methods in automatic extracting." Journal of the ACM (JACM) 16.2 (1969): 264-285.

[Ferr13]   Ferreira R, de Souza Cabral L, Lins RD, de França Pereira E Silva G, Freitas F, Cavalcanti GD, Lima R, Simske S, Favaro L. "Assessing sentence scoring techniques for extractive text summarization," *Expert systems with applications,* 40(14), 5755-5764, 2013.

[Freu95]   Freund Y, Schapire RE, "A decision-theoretic generalization of on-line learning and an application to boosting," *European conference on computational learning theory*, 23-27, Springer, Berlin, Heidelberg, 1995.

[Hoar98]   Hoard JE, "Linguistics and Natural Language Processing," in *Using Computers in Linguistics: A Practical Guide*, ed. by Lawler J, Dry HA, Routledge, 1998.

[Jones93] Jones KS, "What might be in a summary?" Information retrieval 93 (1993): 9-26.

[Kai04]   Kaikhah K, "Automatic text summarization with neural networks," 2004 2nd International IEEE Conference on 'Intelligent Systems'. Proceedings (IEEE Cat. No. 04EX791). Vol. 1. IEEE, 2004.

[Kriz12]   Krizhevsky A, Sutskever I, Hinton GE, "Imagenet classification with deep convolutional neural networks," *Advances in neural information processing systems*, 1097-1105, 2012.

[Kup95]   Kupiec J, Pedersen J, Chen F, "A trainable document summarizer," Proceedings of the 18th annual international ACM SIGIR conference on Research and development in information retrieval, 1995.

[LeCu15]   LeCun Y, Bengio Y, Hinton G, "Deep learning," *Nature* **521**(7553), 436-444, 2015.

[Lin04]   Lin CY, "Rouge: A package for automatic evaluation of summaries," *ACL-04 Workshop*, Association for Computational Linguistics, pp. 74-81, Barcelona, Spain, July 2004.

[Lins19a]   Lins RD, Oliveira H, Cabral L, Batista J, Tenorio B, Ferreira R, Lima R, de França Pereira E Silva, G, Simske SJ, "The cnn-corpus: A large textual corpus for single-document extractive summarization," *Proceedings of the ACM Symposium on Document Engineering* 2019, article 16, 1-10, 2019.

[Lins19b]   Lins RD, Oliveira H, Cabral L, Batista J, Tenorio B, Salcedo DA, Ferreira R, Lima R, de França Pereira E Silva, G, Simske SJ, "The cnn-corpus in Spanish: A large corpus for extractive text summarization in the Spanish language," *Proceedings of the ACM Symposium on Document Engineering* 2019, article 38, 1-4, 2019.

[Lins19c]   Lins RD, Ferreira R, Simske SJ, "DocEng'19 Competition on Extractive Text Summarization," Proceedings of the 2019 ACM Symposium on Document Engineering (DocEng '19), ACM, New York, NY, USA, 216–217.

[Luhn58]   Luhn HP, "The automatic creation of literature abstracts," IBM Journal of research and development 2.2 (1958): 159-165.

[Nall16]   Nallapati R, Zhou B, Gulcehre C, Xiang B, "Abstractive text summarization using sequence-to-sequence rnns and beyond," *arXiv preprint arXiv:1602.06023*, 2016.

[Oliv16]   Oliveira H, Ferreira R, Lima R, Lins RD, Freitas F, Riss M, Simske SJ, "Assessing shallow sentence scoring techniques and combinations for single and multi-document summarization," Expert Systems with Applications **65**, 68-86, 2016.

[Rush71]   Rush JE, Salvador R, Zamora A, "Automatic abstracting and indexing. II. Production of indicative abstracts by application of contextual inference and syntactic coherence criteria," Journal of the American Society for Information Science 22.4 (1971): 260-274.

[Rush15]   Rush AM, Chopra S, Weston JA, "Neural attention model for abstractive sentence summarization," *arXiv preprint arXiv:1509.00685*, 2015.

[Schm15]   Schmidhuber J, "Deep learning in neural networks: An overview," *Neural networks* **61**, 85-117, 2015.

[See17]   See A, Liu PJ, Manning CD, "Get to the point: Summarization with pointer-generator networks," *arXiv preprint arXiv:1704.04368*, 2017.

[Sims13]   Simske SJ, "Meta-Algorithmics: Patterns for Robust, Low-Cost, High-Quality Systems," IEEE Press and Wiley, 386 pp., 2013.

[Sims17a]   Simske SJ, Vans AM, "Using a Large Set of Weak Classifiers for Text Analytics," *Archiving Conference*, Society for Imaging Science and Technology, Vol. 2017, No. 1, pp. 146-151, 2017.

[Sims17b]   Simske SJ, Vans AM, "Learning before learning: reversing validation and training," *Proceedings of the 2017 ACM Symposium on Document Engineering*, 137-140, 2017.

[Sims19]   Simske SJ, "Meta-Analytics: Consensus Approaches and System Patterns for Data Analysis," Elsevier (Morgan Kaufmann), 340 pp., 2019.

[Svor07]   Svore K, Vanderwende L, Burges C, "Enhancing single-document summarization by combining RankNet and third-party sources," Proceedings of the 2007 joint conference on empirical methods in natural language processing and computational natural language learning (EMNLP-CoNLL), 2007.

[Wong08]   Wong K-F, Wu M, Li W, "Extractive summarization using supervised and semi-supervised learning," Proceedings of the 22nd international conference on computational linguistics (Coling 2008), 2008.

3

Clustering, Classification, and Categorization

"Cluster together like stars"
– Henry Miller

"I don't want to be put in a category"
– Winnie Harlow

"There are three kinds of lies: lies, damned lies, and statistics"
– Source unknown

Abstract

Clustering, classification, and categorization are the triad of processes by which unstructured data is turned into information. In a sense, they provide discrimination of information from noise. Clustering associates related content, while classification then assigns meaning to the clusters which can potentially reorganize the clusters. Finally, categorization is the process by which rich data and metadata is associated with the content in each class. This includes tagging, indexing, and assigning document relevance and other statistics. In this chapter, we show how to simultaneously consider these three – clustering, classification, and categorization – to create more robust and complete systems of content. The use of regularization to provide mechanisms for optimizing the information created is highlighted. The relationship between statistical approaches and classification are illustrated.

3.1 Introduction

In this section, we introduce the general characteristics of clustering (Section 3.1.1), followed by an introduction to the very important concept of regularization. Then, we apply regularization at a small scale to clustering. We also argue that, to the extent possible, clustering, classification, and categorization should be performed independently. Regularization tends to work in an opposing fashion, allowing us to combine multiple processes in a weighed fashion. Thus, our preference for independence prevents us from overusing regularization.

3.1.1 Clustering

Humans are good at discrimination. Yes, this is an overloaded statement, as the world's history is unfortunately filled with inequity and partisanship. But, here we mean discriminating one type of information from another, which is a positive form of discrimination, allowing humans to differentiate a king snake from a coral snake, and a gas leak from a crockpot of beans. Discrimination is used to assess situations, to express preferences, and to ensure safety, quality, and consistency. This latter form of discrimination, recognizing and discerning the difference between one thing and another, is no doubt somehow connected to negative aspects, but we are going to focus on the positives in this chapter.

The three types of discrimination we will be concerned with are clustering, classification, and categorization. It is important for us to define how we will use these terms in this chapter and the rest of the book since they are not unambiguous terms. Essentially, we view them as stages of increasing structure and information being added to content as it progresses from unstructured data to fully described information. We begin with clustering. Clustering is the aggregation of data based on relative proximity in some domain. Distance is a common domain, but that is itself a very broad term. Distance can be, for example, linear distance on a map. Suppose we consider a relatively simple one-dimensional (1D) case wherein we try to cluster nine cities that are (just barely) within a day's drive of the starting point (say, Milwaukee) and the end point (say, Grand Forks). Figure 3.1 is a map of the area with stars indicating each city. The distances between these cities as we travel are provided in Table 3.1. We may readily cluster Madison and Milwaukee, Fargo and Grand Forks, and Minneapolis and Saint Paul, and then be left scratching our heads over places such as Fergus Falls, Eau Claire, and Saint Cloud. This is in spite of the fact that Fergus Falls is closer to Fargo than Milwaukee is to Madison, as shown in Table 3.1.

Ideally, we would assign these cities to clusters based on distance alone, and in many cases, that makes sense. However, there is other information to consider. Fargo and Grand Forks, for example, are among North Dakota's largest cities, both located on the Red River on the eastern boundary of North Dakota, and both hosting the state's two large universities. They have a lot in common so that their relative proximity compared to the mean of 264 miles between cities in Table 3.1 is not the only factor involved in forming

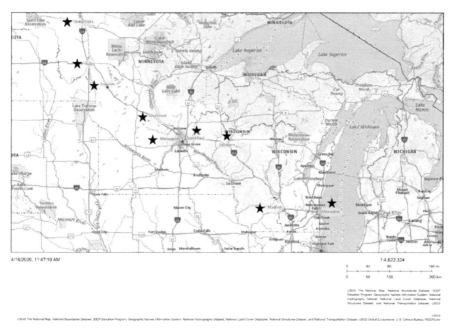

Figure 3.1 Map of the area between Grand Forks, North Dakota, and Milwaukee, Wisconsin [USGS20].

Table 3.1 List of mileages between nine cities in the upper Midwest, USA. Mileages are actually in miles.

	Milwaukee	Madison	Minneapolis	Saint Paul	Fargo	Grand Forks	Fergus Falls	Saint Cloud	Eau Claire
Milwaukee	0	79	337	327	571	651	513	402	246
Madison	79	0	268	259	503	583	445	334	178
Minneapolis	337	268	0	12	235	315	177	66	93
Saint Paul	327	259	12	0	245	324	186	75	85
Fargo	571	503	235	245	0	81	59	176	326
Grand Forks	651	583	315	324	81	0	139	256	406
Fergus Falls	513	445	177	186	59	139	0	118	268
Saint Cloud	402	334	66	75	176	256	118	0	156
Eau Claire	246	178	93	85	326	406	268	156	0

a Grand Forks–Fargo cluster. Fargo is, in mileage, closer to Fergus Falls, Minnesota (59 miles) than it is to Grand Forks (81 miles), but it does seem logical to people from the region that Fargo and Grand Forks are more obviously clustered. It is important, therefore, to list some of the reasons we may override or augment distance (or **proximity**) in deciding on cluster assignment:

(1) Similarity of the elements in the cluster. This generally means contextual information, such as metadata about the items. In the example given, this might mean the size of the cities, their relative location in a state, whether they host a major university, etc.
(2) Membership within a partition or segment that is well-understood. In the context of the example, this may be a state boundary.
(3) Relative proximity. In the context of the above example, this is based on relative distance between locations. For example, Milwaukee and Madison are 79 miles apart, but the next closest city to either of them is Eau Claire, at 178 miles from Madison. Thus, the relative proximity of Madison to Milwaukee is 178/79 = 2.25. This is far more than the Fargo to Fergus Falls compared to Fargo to Grand Forks relative proximity of 1.37.

These three reasons to augment clustering, however, can be viewed as combining classification or categorization rules with those of clustering. For purposes of clustering as an independent activity from that of classification or categorization, then, we should ensure that only the aspects of similarity, membership, and relative proximity that can be **defensibly linked with distance** should be used in the clustering stage of the analysis. We will return to this after investigating how far we can go with just the use of raw distances.

The simplest clustering is one based solely on distance. Table 3.2 shows the clusters that result when a 75-mile distance is deemed the clustering threshold. With this threshold, two clusters are formed: {Minneapolis, Saint Paul, Saint Cloud} and {Fargo, Fergus Falls}. This leaves four cities unassigned to clusters.

From the vantage point of having two small clusters and four unassigned cities, we viewed the clustering to be too conservative. Increasing the distance threshold to 100 miles, which will newly aggregate between 75 and 100 miles, we obtained Table 3.3. Here, every city belongs to a cluster, and no city belongs to two clusters (i.e., two clusters from a previous clustering have not merged).

This seems a good clustering. Next, we increase the clustering threshold to 150 miles, which is more ambitious at clustering cities together. Thanks to the Fergus Falls to Saint Cloud distance of 118 miles, we end up with two clusters only — Madison and Milwaukee in one cluster and the other seven cities in the second cluster. Two clusters from the previous threshold have merged (Table 3.4).

Note that Saint Cloud connects the Twin Cities cluster to Fergus Falls and the North Dakota cluster at a threshold of 118 miles or more. Note that the Wisconsin cluster connects at 79 miles, and Eau Claire connects to the Twin Cities cluster at 85 miles. The optimal cluster range is, thus, based on our intuitive "feeling" for the clustering, the range of (85, 118) miles, the

Table 3.2 Connections to other cities within 75 miles. Clusters are evident from any "X" not along the diagonal of the matrix. For this distance, only two clusters are formed: {Minneapolis, Saint Paul, Saint Cloud} and {Fargo, Fergus Falls}.

	Milwaukee	Madison	Minneapolis	Saint Paul	Fargo	Grand Forks	Fergus Falls	Saint Cloud	Eau Claire
Milwaukee	X								
Madison		X							
Minneapolis			X	X				X	
Saint Paul			X	X				X	
Fargo					X		X		
Grand Forks						X			
Fergus Falls					X		X		
Saint Cloud			X	X				X	
Eau Claire									X

geometric mean of which is 100 miles. Between these two cluster threshold range limits, we have three clusters involving all nine cities, with no previous two legitimate clusters having been merged.

3.1.2 Regularization – An Introduction

How does this "intuitive feeling" translate into an algorithm? As we will see throughout this chapter, it is through regularization. Regularization factors are perhaps best known as algorithmic mechanisms for penalizing overfitting. We will illustrate this using a regression curve. If we have 20 points in a data set, where there is an obvious positive relationship between the independent variable (by custom, the x-axis) and the dependent variable (also by custom, the y-axis), then we may be interested in fitting the data as closely with a model, or regression curve, as possible. Figure 3.2 shows the data set for this example. It is obvious that the dependent variable is positively correlated with the independent variable. But, with what relationship?

The simplest relationship between the independent and dependent variables is a linear regression model, as shown in Figure 3.3. For this model, the output is simply a multiple of the input, minus the offset at zero, if there

Table 3.3 Connections to other cities within 100 miles. Clusters are evident from any "X" not along the diagonal of the matrix. For this distance, three clusters are formed: {Milwaukee, Madison}, {Minneapolis, Saint Paul, Saint Cloud, Eau Claire}, and {Fargo, Grand Forks, Fergus Falls}.

	Milwaukee	Madison	Minneapolis	Saint Paul	Fargo	Grand Forks	Fergus Falls	Saint Cloud	Eau Claire
Milwaukee	X	X							
Madison	X	X							
Minneapolis			X	X				X	X
Saint Paul			X	X				X	X
Fargo					X	X	X		
Grand Forks					X	X			
Fergus Falls					X		X		
Saint Cloud			X	X				X	X
Eau Claire			X	X				X	X

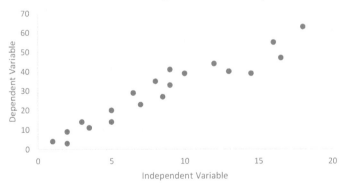

Data Points for Regularization Example

Figure 3.2 Simple regularization example, data points. It is obvious that there is a positive correlation between the independent variable (*x*-axis) and the dependent variable (*y*-axis).

Table 3.4 Connections to other cities within 150 miles. Clusters are evident from any "X" not along the diagonal of the matrix. For this distance, two clusters are formed: {Milwaukee, Madison} and {Minneapolis, Saint Paul, Saint Cloud, Eau Claire, Fargo, Grand Forks, Fergus Falls}.

	Milwaukee	Madison	Minneapolis	Saint Paul	Fargo	Grand Forks	Fergus Falls	Saint Cloud	Eau Claire
Milwaukee	X	X							
Madison	X	X							
Minneapolis			X	X	X	X	X	X	X
Saint Paul			X	X	X	X	X	X	X
Fargo			X	X	X	X	X	X	X
Grand Forks			X	X	X	X	X	X	X
Fergus Falls			X	X	X	X	X	X	X
Saint Cloud			X	X	X	X	X	X	X
Eau Claire			X	X	X	X	X	X	X

is one. For this example, the relationship is not a poor model since 91.25% of all the variation in the dependent variable is explained by changes in the independent variable. This is expressed by the correlation coefficient, R^2, which is equal to 0.9125 for the linear regression model.

This is a good model; however, the linear model has a visibly "flatter" relationship between the independent and dependent variables for values from $x = 9$ to $x = 15$. Here, the slope of the model is the same across the entire input (independent variable) domain. In order to investigate the effect of a non-linear model on this subdomain {9, 15}, successively higher order regressions were performed. The highest order regression to provide a monotonic curve (slope is always positive) across the domain is a sixth-order regression, shown in Figure 3.4. For these data, seventh or higher order regressions have at least one stretch of negative slope, so we terminated the regression at sixth order.

The curve in Figure 3.4 is, of course, a good fit, but there are indications that it may be an overfitted curve. Perhaps, it is too flat between $x = 10$ and $x = 15$. To determine if that is actually happening, though, we are trying to minimize the squared error of the distances between the regression curve and

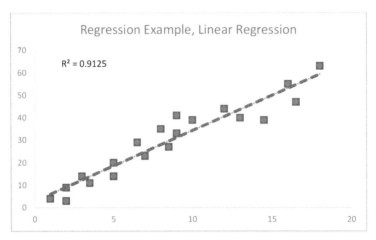

Figure 3.3 Simple regularization example, linear regression (first order). The correlation coefficient, R^2, is equal to 0.9125.

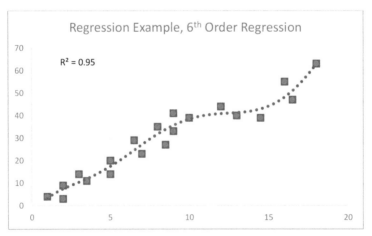

Figure 3.4 Simple regularization example, overfit regression order (sixth order) with monotonicity maintained. The correlation coefficient, R^2, is equal to 0.95.

the empirical data on one hand and the magnitudes of the coefficients (along with the number of coefficients) on the regression equations on the other.

The sum squared errors (SSEs) for the residuals of the regression and the original values is defined from: $\mathrm{SSE} = \sum_{i=1}^{n_{\mathrm{samples}}} (y_i - \hat{y}_i)^2$ where $\hat{y}$ is the regression value and y is the empirical value (measured). To this SSE, which we wish to minimize, we add a penalty for the (scalar) magnitude of the coefficients and, additionally, the number of coefficients. This penalty,

Table 3.5 Regression equations and norm of the regression equation coefficients for the first-, third-, and sixth-order regressions.

Order	Regression equation	Norm $\|w\|$
1	$y = 3.1377x + 2.9083$	4.28
3	$y = 0.0128x^3 - 0.4286x^2 + 7.1562x - 6.0697$	9.39
6	$y = -7\text{E-}05x^6 + 0.0044x^5 - 0.0984x^4 + 0.9923x^3$ $- 4.5922x^2 + 12.567x - 5.7545$	14.60

Table 3.6 SSE, standard error of the mean (SEM), F-statistic, and p-value of the residuals for the first-, third-, and sixth-order regression models are given in columns 2−4. The cost function, $J = $ SSE $+ 11.23(\|w\|+n_{\text{coefficients}})$, given by Equation (3.3), is in the last column.

Order	Residual SSE	Residual SEM	Residual F-statistic (p)	J (Equation (3.3))
1	480.50	5.167	187.6 (5.8×10^{-11})	551.02
3	387.62	4.922	70.2 (2.0×10^{-9})	537.99
6	274.27	4.593	41.2 (1.1×10^{-7})	516.84

then, is typically in the form of $\lambda(\|w\| + n_{\text{coefficients}})$, where $\|w\|$ is the norm of the coefficients (or "weights") of the regression equation given by $\hat{y} = \sum_{j=0}^{J} w_j x^j$. Because the weights of the coefficients (along with the number of coefficients used) are penalized during optimization, they are prevented from growing to fill in all the degrees of freedom. The overall cost function, J, is SSE$+ \lambda \|w\|$, or $J = \sum_{i=1}^{n_{\text{samples}}} (y_i - \hat{y}_i)^2 + \lambda(\|w\|+n_{\text{coefficients}})$. We first need to set the regularization factor λ. We do this such that for the simplest model, the linear regression, the contribution from the SSE of the residuals from the regression, or $\sum_{i=1}^{n_{\text{samples}}} (y_i - \hat{y}_i)^2$ is equal to the contribution from the norm, or $\lambda \|w\|$, multiplied by the minimum degrees of freedom in the number of samples, which we set to $(n_{\text{samples}}/2)$ since we do not allow the order of the regression to be more than half the number of samples; that is:

$$\sum_{i=1}^{n_{\text{samples}}} (y_i - \hat{y}_i)^2 = \lambda \|w\| (n_{\text{samples}}/2). \tag{3.1}$$

In order to calculate λ, we must first know the values of $\|w\|$ for each of the regression equations of interest. These are given in Table 3.5, where for linear regression, we see that $\|w\| = 4.28$. The statistics for the residuals are given in Table 3.6.

For the linear regression case, SSE = 480.50, and $\|w\| = 4.28$. Since $n = 20$, we use minimum df = $n/2 = 10$, and we have:

$$480.50 = (10)\lambda(4.28). \tag{3.2}$$

From this, $\lambda = 11.23$. Our overall cost function, J, is thus given by Equation (3.3) for every order of regression, where $n_{\text{coefficients}} = 2, 4$, or 7, respectively, for the first-, third-, and sixth-order regressions:

$$J = \sum_{i=1}^{n_{\text{samples}}} (y_i - \hat{y}_i)^2 + 11.23(\|w\| + n_{\text{coefficients}}). \qquad (3.3)$$

From the results presented in Table 3.6, we decide that the sixth-order regression, in spite of the penalty for higher coefficient weights and higher number of coefficients, is still a better fit for the data than either the linear regression model (Figure 3.3) or the third-order regression (Figure 3.5). After regularization, the fitness of these three regression models is much more similar (a 6.6% difference between the first- and sixth-order regressions) than solely based on SSE of the residuals (a 75.2% difference between the first- and sixth-order regressions). While there are other ways to regularize these expressions (please see, for example, Chapter 10 of reference [Sims19]), these two regularization approaches illustrate its operation and value for relatively comparing different data models.

The code to create these curves in R software is relatively straightforward. For simplicity of the reader using the code, it is included in Figure 3.6. It can be readily modified to include the regularization in addition to producing regression models of different orders from the first, third, and sixth orders included in this discussion. Regardless, this subsection introduces the reader to the regularization process, which allows us to optimize models by trading off the error in the model with the expense of defining the model. In any clustering, classification, or categorization approach, we need to be able to trade off the goodness of fit with prevention of overfitting. Regularization approaches provide that balance, as we consider next by returning our attention to the original clustering example of the previous subsection.

3.1.3 Regularization and Clustering

Returning to the simple clustering example, with the perspective of regularization, we can now "penalize" clustering with factors other than distance. The first thing to note is that our earlier exploration of the clustering for the nine cities was based solely on distance and ended up with a "sweet spot" for assigning each city to exactly one cluster: a distance threshold range of (85, 118) miles. But, how does this range relate to a relevant statistical measurement, such as the analysis of variance (ANOVA), its associated F-scores, and the p-values for statistical significance of the ANOVA output? In order to explore that, we re-evaluated this sample set of nine midwestern US cities using the 75, 100, and 150 mile thresholds from Section 3.1.1 and, in addition, added a 200-mile threshold. The R code corresponding to this investigation is given in Figure 3.7. The code is used to assess the

Figure 3.5 Simple regularization example, optimum regression order (third order) after regularization. The correlation coefficient, R^2, is equal to 0.9294.

```
regressions_test
{
        x = c(1,2,2,3,3.5,5,5,6.5,7,8,8.5,9,9,10,12,13,14.5,16,16.5,18)
        y = c(4,9,3,14,11,20,14,29,23,35,27,41,33,39,44,40,39,55,47,63)
        model1 = lm(y ~ poly(x,1))
        res1 = resid(model1)
        sum1 = 0
        for(i in seq_along(res1) )
                sum1 = sum1 + (res1[i]*res1[i])
        model3 = lm(y ~ poly(x,3))
        res3 = resid(model3)
        sum3 = 0
        for(i in seq_along(res3))
                sum3 = sum3 + (res3[i]*res3[i])
        model6 = lm(y ~ poly(x,6))
        res6 = resid(model6)
        sum6 = 0
        for(i in seq_along(res6) )
                sum6 = sum6 + (res6[i]*res6[i])
        print(sum1)
        print(sum3)
        print(sum6)
        summary(model1)
        summary(model3)
        summary(model6)
}
```

Figure 3.6 Simple *R* code to calculate the regression values shown in Figures 3.2–3.5 and Tables 3.5 and 3.6. Only first-, third-, and sixth-order regression models are calculated here. Output from the summary() statements provides the data given in Tables 3.5 and 3.6.

```
clustering_anova_tests
{
        # 75 miles for a cluster
        g1 = c(12,59,66,75)
        g2                                                                =
c(79,337,268,327,259,571,503,235,245,651,583,315,324,81,513,445,177,186,139,402,334,176,2
56,118,246,178,93,85,326,406,268,156)
        aov1 = c(g1,g2)
        labels1 = c(rep("g1",4),rep("g2",32))
        fit1 = aov(aov1~labels1)
        summary(fit1)

        # 100 miles for a cluster
        g3 = c(12,59,66,75,79,81,93,85)
        g4                                                                =
c(337,268,327,259,571,503,235,245,651,583,315,324,513,445,177,186,139,402,334,176,256,118
,246,178,326,406,268,156)
        aov2 = c(g3,g4)
        labels2 = c(rep("g3",8),rep("g4",28))
        fit2 = aov(aov2~labels2)
        summary(fit2)

        # 150 miles for a cluster
        g5 = c(12,59,66,75,79,81,93,85,139,118)
        g6                                                                =
c(337,268,327,259,571,503,235,245,651,583,315,324,513,445,177,186,402,334,176,256,246,178
,326,406,268,156)
        aov3 = c(g5,g6)
        labels3 = c(rep("g5",10),rep("g6",26))
        fit3 = aov(aov3~labels3)
        summary(fit3)

        # 200 miles for a cluster
        g7 = c(12,59,66,75,79,81,93,85,139,118,177,186,176,178,156)
        g8                                                                =
c(337,268,327,259,571,503,235,245,651,583,315,324,513,445,402,334,256,246,326,406,268)
        aov4 = c(g7,g8)
        labels4 = c(rep("g5",15),rep("g6",21))
        fit4 = aov(aov4~labels4)
        summary(fit4)

        # summary() output
                Df Sum Sq Mean Sq F value Pr(>F)
        labels1    1 199817 199817   9.061 0.0049 **
        Residuals  34 749788  22053
        labels2    1 391003 391003   23.8 2.47e-05 ***
        Residuals  34 558602  16429
        labels3    1 463806 463806   32.46 2.13e-06 ***
        Residuals  34 485799  14288
        labels4    1 591933 591933   56.27 1.05e-08 ***
        Residuals  34 357672  10520
}
```

Figure 3.7   Simple *R* code to calculate the cluster-related ANOVA output as per Section 3.1.3. Four different distance thresholds were evaluated: 75, 100, 150, and 200 miles. The *p*-values for the *F*-scores for these four distances were 4.9×10^{-3}, 2.47×10^{-5}, 2.13×10^{-6}, and 1.05×10^{-8}, respectively. From a ratio of between cluster to within cluster variability perspective, the 200-mile threshold was the best of the four investigated.

ANOVA, or analysis of variance (we will see much more on ANOVA later in this chapter), of the clusters formed. The "between cluster" values are the distances between the mean locations of the clusters, whilst the "within cluster" values are the mean distances between cities belonging to a single cluster. From the **summary() output** statements at the bottom of the code shown in Figure 3.7, we can see that the p-values for the F-scores for cluster distance thresholds of 75,100,150, and 200 miles are 4.9×10^{-3}, 2.47×10^{-5}, 2.13×10^{-6}, and 1.05×10^{-8}, respectively. These correspond to F-scores of 9.06, 23.8, 32.5, and 56.3, respectively.

Our general approach to regularization is to use a regularization equation of the form given in Equation (3.4):

$$J = p\text{-value}(F\text{-score(clustering)}) + \lambda_1(\%\text{combined}) + \lambda_2(\%\text{unclustered}).$$
$$(3.4)$$

From one iteration of clustering to the next, we penalize two factors in Equation (3.4). The first is **%combined**, which is the percentage of all cities that were in separate clusters previously that have now been combined into a larger cluster. For example, when moving the distance threshold from 100 to 125 miles, the Twin Cities (Minneapolis–Saint Paul) and North Dakota clusters combine, meaning three of the nine cities, or 0.333 of the cities, are now combined. In moving from 75 to 100 miles, no previous clusters were combined, and so this value, **%combined**, is 0.000. The second factor that can be penalized is the **%unclustered**, the percent of cities not currently assigned to a cluster. The values for these for 75, 100, 150, and 200 miles are given in Table 3.7.

As Table 3.7 shows, the overall objective (or "J") function is minimized when the threshold for clustering is 100 miles (range $85-118$). Here, we used $\lambda_1 = \lambda_2 = 0.01$ for the two regularization factors. These were chosen to be 0.01 since the p-values of the F-scores were all below 0.005. However, the results are not sensitive to the value of the regularization coefficients above 0.0001; for every such case, the clusters formed with a distance threshold of 100 are rated the optimum (have the minimum J value in the last column of Table 3.7).

As mentioned above, there are three "legitimate" instances in which we can augment distance (or "proximity") in deciding on cluster assignment, without stepping into the territory normally associated with classification or categorization (see following sections for more on those). These are:

(1) Similarity of the elements in the cluster: in the context of geographical/map clustering, this might mean the size of the cities, their relative location in a state, whether they host a major university, etc.
(2) Membership within a partition or segment that is well-understood: in the context. In the context of the example, this may be a state boundary.

Table 3.7   The p-value of the F-scores for clustering thresholds of 75, 100, 150, and 200 miles, as per the example in Section 3.1.1. $J = p$-value(F-score(clustering)) + λ_1(%combined) ⊢ λ_2(%unclustered), as per Equation (3.4), is given in the last column, and is based on %combined and %unclustered in columns 3 and 4. For all values of λ_1 and λ_2 greater than 0.000075, the optimum J is the one for Threshold = 100.

Threshold	p-value(F-score)	%combined	%unclustered	$J\|\lambda_1 = \lambda_2 = 0.01$
75	0.0049	0.000	0.333	0.004900
100	2.47e-05	0.000	0.000	0.000025
150	2.13e-06	0.333	0.000	0.003335
200	1.05e-08	0.333	0.000	0.003333

(3) Relative proximity. In the context of the above example, this is based on relative distance between locations. For example, Milwaukee and Madison are 79 miles apart, but the next closest city to either of them is Eau Claire, at 178 miles from Madison. Thus, the relative proximity of Madison to Milwaukee is 178/79 = 2.25. This is far more than the Fargo to Fergus Falls compared to Fargo to Grand Forks relative proximity of 1.37.

Even for these three reasons to augment clustering, we must be careful to keep clustering as an independent activity from that of classification or categorization: only the aspects of similarity, membership, and relative proximity that can be **definitively linked with distance** should be used in the clustering stage of the analysis. Given the rivalries between states, adding the equivalent of an hour's driving (60 miles) to crossing each state boundary is a defensible clustering addition. When we do this (Table 3.8), we first form a set of clusters involving all nine cities for a clustering distance threshold of 118 miles, and the clusters are {Milwaukee, Madison}, {Eau Claire, Saint Paul, Minneapolis, Saint Cloud, Fergus Falls}, and {Fargo, Grand Forks}. With the exception of Eau Claire, these clusters now precisely match state boundaries, which makes more sense, intuitively, than many other possibilities. Besides, as noted in point (3), from a relative proximity standpoint Eau Claire is a lot closer to the Twin Cities (Minneapolis and Saint Paul) than it is to the other two cities in Wisconsin (Milwaukee and Madison).

With this introduction to clustering, regularization, and the application of regularization to clustering, we move to the more general consideration of clustering, classification, and categorization.

Table 3.8 List of weighted mileages between nine cities in the upper Midwest, USA. Mileages are actual miles between cities, +60 miles for every state border crossed between cities. Here, the first clustering that includes all nine cities in a cluster is at a cluster distance threshold of 118 miles, for which the clusters are {Milwaukee, Madison}, {Eau Claire, Saint Paul, Minneapolis, Saint Cloud, Fergus Falls}, and {Fargo, Grand Forks}.

	Milwaukee	Madison	Minneapolis	Saint Paul	Fargo	Grand Forks	Fergus Falls	Saint Cloud	Eau Claire
Milwaukee	0	79	397	387	691	771	573	462	246
Madison	79	0	328	319	623	703	505	394	178
Minneapolis	397	328	0	12	295	375	177	66	153
Saint Paul	387	319	12	0	305	384	186	75	145
Fargo	691	623	295	305	0	81	119	236	446
Grand Forks	771	703	375	384	81	0	199	316	526
Fergus Falls	573	505	177	186	119	199	0	118	328
Saint Cloud	462	394	66	75	236	316	118	0	216
Eau Claire	246	178	153	145	446	526	328	216	0

3.2 General Considerations

In moving from clustering to classification and then to categorization, we also move from definition to description and then to knowledge generation. This analogy drives the meanings of each of these three terms in our text analytics system and, as such, serves as the guide for where to position each of our algorithms. Clustering, as the definition phase, **provides boundaries for and between different elements of data**. In order to enable "functional" analysis of the text, in general, the clustering approach is performed in a state of ignorance and uses only rules of proximity (referred to as distance in the section above), similarity, membership, and relative proximity to guide the assignment of information to the clusters.

The central task of intelligent text analytics is classification, and, as such, it is the easiest task to conflate with the other two tasks. Classification that is overly concerned with the cluster boundaries may limit itself to terms only evident in the training data, which will handicap its later ability to learn. On the other side, classification that is overly concerned with categorization may avoid a validation step when it is sorely needed and move directly to defining the expected contents of any indexing, tagging, or labeling (collectively, "categorization") to follow the classification and thus provide an overfitted solution to the text analytical challenge. Classification **provides belonging to content**, which is often highly correlated with boundaries, but may proceed from a different starting point. For example, clustering is often the means to add boundary-based structure to unstructured content, while classification is the way to take existing structure, in the form of a set of relevant classes to the end user or end application, and retrofit the clustered data to be assignable to the classes. **Classification fits a structure onto data that has been independently assigned to clusters**. A class structure may or may not be in balance with the clustering or the relative composition of the training data. This means that classification is most concerned with the training, validation, and test sets and must use an experimental design best architected to provide stable classification when the system is deployed.

Categorization is often defined in reference to classification. Categorization is the means of assigning knowledge generated in the process of clustering and classification to the content that is now presumably more effectively aggregated and structured than before. Categorization is **knowledge generation concerned with the indexing, tagging, or labeling of classes such that they can be employed quantitatively** in useful analytical tasks by users, applications, and services derived from the content. Without categorization, we have a relatively static and thus largely qualitative description of the content. With categorization, we have differential significance assigned to different elements of the content. Categorization is thus the process by which learning occurs, and our content has moved from perhaps fully unstructured content to content that is fully contextual.

Thinking about this in broad and tangible human terms, then, clustering is a subset of content, perhaps a chapter in a book or an article in a magazine. Classification is then the subject matter of the book – perhaps a novel about the Troubles in Northern Ireland by Anna Burns, for example – or the magazine – this week's Economist, for instance. Categorization is then the subject matter of the specific section of text – paragraph, article, etc. – and gives us the learning elements from that particular content. As we move from clustering to classification, we increase the level of understanding – and thus perform learning – by elucidating some of the reasons why the clusters occurred in the first place. A similar perspective links categorization to classification: we learn more about why the classification took place when we index, tag, or, otherwise, label the classes. Importantly, this three-step process, if appropriately applied, mirrors the three-step process of training, validation, and testing that accompanies any reasonable system design in the first place. Not to stretch the analogy too far, but, effectively, a clustering of the content is the training phase for overall system learning generation: it is a suggestion, based on factors such as proximity, similarity, membership, and relative proximity, of how the content may be structured. Then, classification is the validation phase, associating elements of a cluster with a choice of class assignment and comparing the (preferably independent) assignment of class with the cluster membership integrity. Categorization, then, is the test case, wherein the consistency of labeling across a class is used to determine class cohesiveness at the collective level and the breadth of each class by determining the label at the individual element level.

We started this chapter talking about how humans are good at discrimination. Assuming one of the goals of text analytics is to create categorization that is useful to humans, then we may wish to determine whether or not each stage – clustering, classification, and categorization – is in some ways analogous to the human content processing approaches. Clustering is analogous to the skimming of content humans use to assess at a high level the general topic(s) of the content. Here, the topic is assessed from the bottom up; that is, the content and its scanning are used to assess its relative belonging to other content. Classification is then used to assign one or more clusters to a top-down definition of content topic(s). Combining the two, we have the eminently trainable architecture of bottom-up plus top-down to drive the decision of how to associate content. Once associated, we then mine the specific contents of each assigned element and element set (or class) to create the tags, labels, and indices that are the hallmark of the knowledge we generate from the content.

3.3 Machine Learning Aspects

Clustering, classification, and categorization are integrally related to machine learning. In this section, we consider how machine learning approaches can be brought to bear on each of these processes. More importantly, perhaps, we introduce several key statistical and tabular approaches to optimization that are significant for anyone involved in analytics.

3.3.1 Machine Learning and Clustering

In many ways, clustering is simply a partitioning of the variability in data, with constraints. Classic means of determining clusters include the expectation−maximization (EM) approach [Demp77], wherein in the simplest form, there are any number of iterations, and each iteration involves two complementary steps:

(1) Assignment − each item is assigned to the cluster that is closest to it in distance.
(2) Redefinition − each cluster is now redefined based on the samples belonging to it.

The key parameter in such an algorithm is the number of clusters to choose. This is where both statistical approaches and regularization approaches come into play. Statistical approaches to clustering are analogous to the ANOVA, as mentioned above, wherein the goal is to provide as high a ratio as possible of the variance between the centroids of each cluster divided by the variance within the clusters themselves. This is typically called the F-score for ANOVA, given by Equation (3.5):

$$F = \frac{\mathrm{MSE\,(between)}}{\mathrm{MSE\,(within)}}. \tag{3.5}$$

Generally, the p-value of the F-score from Equation (3.5) is used for comparative purposes, but the basic premise is that the greater the relative distance between putative aggregates (or "clusters") is compared to the distances between members of the distinct aggregates, the more confident we are that these aggregates provide an optimal partitioning of the elements. This is intuitive and is as much a part of our language as it is our algorithms. For example, let us consider astronomy, wherein the terminology is directly derived from the clustering of the astronomical bodies. A solar system is based on its centroid, or sun, and the elements of the cluster are the planets, their satellites, and other objects such as asteroids, meteors, interplanetary dust clouds, and comets which are gravitationally bound to the star. The space between solar systems is called interstellar space, and when two or more solar systems are close enough together, they are called a multiple star system. In other words, when the between star distance is relatively small compared to the within solar system distance, they are reclustered and considered a single system.

Since we are concerned with the **functional** aspects of text analytics in this book, we need to consider not just the relative distances involved in separating different clusters but also the functional distances. In our earlier example involving the clustering of nine different cities, for example, clustering cities by state rather than solely by distance makes sense. This is particularly the case in, as one example, an atlas in which the city maps generally accompany the state maps rather than being organized by proximity. In such a circumstance, the clustering is done *by context*.

In addition to statistical approaches, clustering can also be tied to training, validation, and testing results; that is, to accuracy, precision, and recall methods. If, for example, a k-fold cross-validation approach is being used, then only the clusters that are in common across the set of validations are used in the first step (the "tessellation" phase) of the cluster formation. This is shown in Table 3.9. Here, four separate clustering attempts were performed. The clustering algorithm employed for this example was simply the random selection of 4 of the 13 samples (labeled A through M) as the seeds of the four clusters and then assigning the rest of the samples to the sample closest to them (admittedly not the best clustering approach, but, for this example, it provided a diversity of clustering results).

Tessellation of the fourfold clustering allows clusters to be maintained only where the samples are part of the same cluster for all four procedures. In the example shown in Table 3.9, the three elements of the triplet $\{A, B, C\}$

Table 3.9 k-fold cross-validation where $k = 4$. Individual clusters for each of the four validation sets are given in {} above. The "Tessellation" of the clusters is given in the bottom row.

Fold number	Clusters of samples A—M
1	{A, B, C}, {D, E, F, G, H}, {I, J, K}, {L, M}
2	{A, B, C, D}, {E, F, G}, {H, I, J, K}, {L, M}
3	{A, B, C, E}, {D, F, G, H}, {I, J, K, L}, {M}
4	{A, B, C}, {D, F, G, H, I}, {E, J, K}, {L, M}
Tessellation	{A, B, C}, {D}, {E}, {F, G}, {H}, {I}, {J, K}, {L}, {M}
Recombination	{A, B, C}, {D, E, F, G, H}, {I, J, K}, {L, M}

always belong to the same cluster. The same is true for {F, G} and {J, K}. These are the "starting" three clusters for the "recombination" step in the process. Next, the unassigned elements – that is, D, E, H, I, L, and M – need to be assigned to either one of these three clusters or else to form the missing fourth cluster. This step can be performed in a number of ways. For our decision, since {L,M} occur together in three of the four cross-validations, they are assigned to the same cluster, which happens to be the new cluster. Then, D, E, H, and I are readily assigned to the clusters to which they most commonly join in the four folds. This tessellation and recombination pattern, originally introduced in [Sims13], is a good means of redefinition, akin to the second part of each iteration in an EM approach.

3.3.2 Machine Learning and Classification

In many cases, when we think of classification, we are especially concerned with the triad of training, validation, and testing. Training is generally the stage when we use a set of labeled (e.g., clustered) data to establish the type of model we wish to use for later deployment of the classifier. Validation is often the stage at which we define the settings for the type of model selected during training. Finally, testing is used to give us a reliable prediction of how well the machine learning will behave after deployment.

In our experience [Sims19a], the experimental design is often the most crucial part of the machine learning/classification process. This is pithily summarized by the comment that "your system is only as good as the data on which it has been trained," which is certainly true. But, this is actually a shallow statement that does not dive into the complexities involved in the experimental design. For instance, if the training phase is meant to allow you to relatively compare the suitability of two or more machine learning approaches (for example, a Bayesian vs. a neural network vs. a support vector approach), and then the actual settings for the chosen approach need to be discerned, this argues for a balanced $1/3 : 1/3 : 1/3$

design in which training, validation, and testing groups are of equal size. By settings, we mean, for example, coefficients and weighting values used by the chosen approach. Consider a Bayesian approach in which we find that the conditional probabilities for the training set are $\{0.60, 0.70, 0.80\}$. Then, for the separate validation step, we allow these values to be modified to obtain the best machine learning results on the validation data and obtain $\{0.65, 0.75, 0.70\}$ as our new conditional probabilities.

A quite different approach might be used when the settings are integrally part of the training process. Such a condition is encountered, for example, in many ensemble approaches, including boosting, stacking, bagging, and meta-algorithmics [Sims13]. In such approaches, the coefficients (weightings, probabilities, etc.) are intentionally updated by the validation data in order to optimize the performance of the system so trained. Meta-algorithmic systems typically incorporate two or more already-trained individual approaches and, thus, evaluate their contribution to the ensemble system on the separate validation data. The lines between training and validation, however, can be readily blurred. In the example of the Bayesian approach, why not just weight the training and validation data equally to obtain, for example, deployment-ready weights of $\{[0.60+0.65]/2, [0.70+0.75]/2, [0.80+0.70]/2\}$, or $\{0.625, 0.725, 0.75\}$? It seems justified on the grounds of using as much data as possible to determine the system settings to simply combine training and validation sets in such a case. If this statement can be made, then, we argue for the merging of validation and training sets: *the validation output provides no more than a simple additional training instance in which different results are possible, but unpredictable in terms of their directional change from the training set.* Under such circumstances, we argue that there is, in fact, no such thing as a validation step, and we are simply using validation to enhance the amount of training data. If this is the case, instead of a $1/3 : 1/3 : 1/3$ design in which training, validation, and testing groups are of equal size, we should employ a $1/2 : 1/2$ design in which training and testing groups are of equal size.

There are, on the other hand, approaches in which validation truly provides a separate, additional insight from the training results. For example, the training data is used to relatively rank a set of approaches for an ensemble method and finds that ensemble methods A, B, C, D, and E are ranked $1-5$ in order for selection by the ensemble method, with relative weights of $\{0.41, 0.24, 0.21, 0.08, 0.06\}$. Next, the validation set is used to provide weights for these five methods, and it shows that D and E have negative impact on system accuracy such that only methods A, B, and C should be part of the deployment set. After the validation step, then, we can assign the weights $\{0.51, 0.27, 0.22\}$ to A, B, and C (the results from the validation phase after D and E are dropped) or combine the weights with the relative weights from the training phase {0.41/0.86, 0.24/0.86, 0.21/0.86} with D and E left out; thus, A, B, and C weights are {0.48, 0.28, 0.24} for the training data. The mean of these two — {0.51, 0.27, 0.22} and {0.48, 0.28, 0.24} — yields an output from training and validation that is {0.495, 0.275, 0.23} and excludes methods D and E. In this example, the validation step was used to prune two of the approaches and was still used in combination with the training set to define the final deployment coefficients; that is, {0.495, 0.275, 0.23}.

In the end, then, validation steps can be viewed as supporting training, testing, or both. If validation is used to follow up the training and determine the settings, it is effectively part of training. If it is used as a separate assessment of system performance, it is effectively a test. Note that in both of these examples, the testing set (which is always set aside from the training and validation set, and not used until the testing stage) is used to report the likely deployment performance of the training + validation output design.

The point of contrasting the two approaches in this section is to show that there is some art involved in the creation of the appropriate experimental design. We emphasize this point by considering the two experimental extremes. The first is the k-fold evaluation, in which the labeled data is partitioned into k equally sized sets, and $k - 1$ sets are used for training with the single "fold" left out used for testing. The testing accuracy is the mean of the accuracies on the k folds left out. This approach leads, innately, to overtraining and, thus, should not be used to estimate the later performance of the approach on data during deployment. The main purpose of k-fold evaluation, or cross-validation as it is also called, is to provide a means for relative comparison of one algorithm to another. It is not a proper way to absolutely assess an approach, although it can be an appropriate way to relatively assess different approaches. The second extreme is the $1/2$ training, $1/2$ testing approach mentioned in this section. This is an optimal experimental design in order to predict how well a system is going to behave when deployed. A 50:50 split of the data for training and testing in general results in the least sensitive predictor of future performance of any experimental design, so long as samples are truly randomly assigned to training and testing groups.

Much more attention will be paid to the role of machine learning in classification later in this chapter (in particular, Section 3.5); however, it is worth noting here that the approach to experimental design is not conceptually dissimilar from the regularization-based approaches that are employed for the optimization of classification in Section 3.5. Effectively, we set up the experimental design to fit the form of Equation (3.6):

$$\text{Design} = (1 - \lambda)(\text{Training}) + \lambda(\text{Testing}) \tag{3.6}$$

In a minimally sensitive (or maximally robust) system design, we set the regularization weight, λ, to 0.5 in Equation (3.6). However, depending on the extent to which our validation data complements the training or the testing approach, we may move λ up in value (when validation is more generally associated with training) or down in value (when validation is akin to a testing experiment).

3.3.3 Machine Learning and Categorization

Categorization typically builds on classification. In many ways, it is the achievement of structure at the end of a long progression from unstructured data to cluster to class to category. In our definition, we view category as being associated with tagging, labeling, and/or indexing. Thus, a category is distinguished from a class, in that it outlines distinctive elements of membership. These elements are distinctive, not discriminating. They need not be features that discriminate one class from another. Instead, they are ones that are integral to the definition of the category. Thus, we may have two classes such as "New Mexico cities" and "Nevada cities" for each of which the element "Las Vegas" is a categorical term but for which classification terms do not include "Las Vegas" but instead "Albuquerque," "Santa Fe," "Reno," and "Sparks." In many ways, classification and categorization, together, are the logical outcome of both machine learning and data mining. The terms associated with classification are the machine learning terms minus those intersecting with the data mining terms; the terms associated with categorization are the union. We term this the *classification−categorization ledger approach*.

For the optimum functional performance of a text-analytic-based system over time, the classification−categorization ledger approach has the advantage of being readily reassigned one term at a time. Thus, when additional documents are to be assigned, they need only be mined for terms associated with the ledger, and these terms updated. On occasion, terms may be moved from one role (e.g., classification) to the other (e.g., categorization).

3.4 Design/System Considerations

Design and system considerations for clustering, classification, and categorization should, if possible, consider the single most important design decision first. Afterwards, as possible, additional decisions can be made. The first decision to be made, however, depends on the reference point. If we have a set of classes already defined, for example, then we likely have some flexibility as far as defining clusters. Do we draw gerrymandered boundaries around the individual classes, or do we allow classes to comprise two or more clusters, which within the dimensions of the problem space are separated by either the bodies of other clusters or even open space? Such multiple-cluster classes can be thought of as aggregate classes, and these classes should be continually re-evaluated to see if they are properly defined as multiple classes later in time. Alternatively, it may be observed that the clusters grow together and so the class boundaries of the two clusters can merge. Regardless, starting with the classes and then performing cluster life-cycle management (e.g., birth, merge, split, and death of individual clusters) is a form of addressing *structured data* since the classes exist before the clusters and presumably need to be maintained irrespective of how the clusters are altered over time. This may result in situations in which, for the purposes of maintaining the integrity of data classes, we choose to ignore some clusters entirely, either because they are relatively small, relative isolated, or relatively unrelated to any clusters in any of the defined classes. Previous work [Sims17] has shown that in cases such as these, it is likely a good practice to ignore some of the training data when defining the settings for the rest of the intelligent system since, otherwise, this training data may lead the system to depend too heavily on these atypical clusters. This approach has been termed "pre-validation" since the removal of parts of the training data from participating in the definition of system settings effectively uses the omission of the data as a form of validation for the use of the rest of the training data.

If, instead of having structured data, we handle *unstructured data*, then we begin with the formation of clusters and thereafter provide the formation of classes. Here, the strict need for a 1:1 correspondence between cluster and class is not encouraged either. Rather, we can let an optimization approach determine automatically the relative degree of clustering for the data. Optimization in the systemic formation of clusters and classes is governed in a straightforward fashion by an equation with one or more constraints imposed. We use regularization approaches to govern the manner in which these constraints are mapped to the equation. Consistent with the use of Langrangian multipliers in differential equations, we multiply the constraint by a factor lambda, which can be set experimentally.

This is illustrated by creating an optimization equation based on the statistical confidence value, or *p*-value, associated with the *F*-score for the clustering, added to the error rate of the classification approach. The simplest form of this approach is given in Equation (3.7):

$$\text{Minimize}: \text{ Error rate} + p(F - \text{score}). \tag{3.7}$$

Here, we simply want the minimum sum of the error rate added to the *p*-value of the *F*-score. Since each of these factors is always greater than or equal to 0.0 and is "optimized" when minimized – that is, exactly equal to 0.0 –

it may seem like Equation (3.7) requires no further consideration. However, since the relationship between error rate and p-value of the F-score is not necessarily consistent across different data sets, we prefer the more general Equation (3.8):

$$\text{Minimize}: \ (1 - \lambda)(\text{Error rate}) + \lambda(p(F - \text{score})). \qquad (3.8)$$

Here, the regularization factor, λ, is multiplied by the p-value of the F-score, shorthanded $p(F$-score) in Equations (3.7) and (3.8). The *Error rate*, meanwhile, is multiplied by $1 - \lambda$. This prevents the overall sum of the expression that is to be minimized from ever exceeding the larger of the two factors — that is, *Error rate* and $p(F$-score) — in case different systems may be compared. Error rate is limited to the range from 0.00 (a perfect system) to 1.0 (a system in which every output is in error), and, similarly, $p(F$-score) is limited to the range from 0.0 to 1.0. This is consistent with the value of the sum in Equation (3.8) itself; for example, if Error rate is high and λ is close to 0.0, then $(1 - \lambda)$ is close to 1.0 and the left side of the equation is nearly 1.0. Regardless, the sum in Equation (3.8) is limited to the range 0.0–1.0.

Equation (3.8) forms the basis of a novel "self-regularizing" means of clustering information. The optimal cluster set has the minimum sum of these two elements. This allows us to simultaneously consider the impact of classification errors with the impact of loosening/tightening the clustering. The latter determines whether we combine two or more clusters into a single larger cluster or, conversely, split a larger cluster into two or more smaller ones. Even simply viewing the relationship between clustering and class assignment as dynamic changes the nature of the analytics approach. Classes that involve multiple manifolds or otherwise comprise multiple clusters can evolve over time from a collective of subclasses into a single primary class or multiple classes with all but one newly created. This type of "preadaptation" to advantageous reclustering is part of robust, resilient, responsive, and responsible system design.

Optimizing a distribution for class and cluster simultaneously is shown through example (Figures 3.8–3.11), with the associated statistics. Figure 3.8 illustrates data that comes from two distinct classes (represented by diamonds and squares). A third class (represented by circles) consists of two potential subclasses, in lighter or darker shade. We have several tasks to complete for this data set, including the following:

(1) Determining whether there should be three or four classes — should the circles be split into two classes or left as one?
(2) Determining whether the diamond and square classes should be represented as single or double clusters.
(3) Determining which combination of classes (3 or 4) and clusters (3–6) to keep.

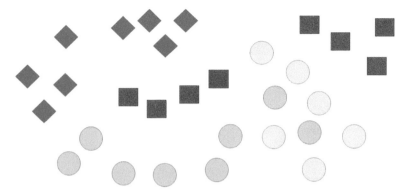

Figure 3.8 Set of data belonging to three classes (diamond, square, and circular elements). The class represented by the circles does not have any obvious subclusters that result in no classification errors since the two potential subclasses of data (lighter and darker filled circles) have extensive overlap.

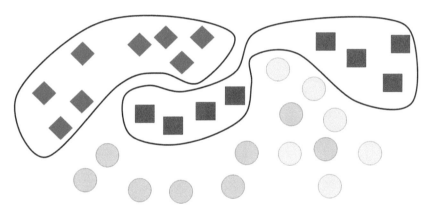

Figure 3.9 Set of data belonging to three classes (diamond, square, and circular elements) with the diamond and square element classes shown as a single cluster. While this arrangement (assuming all of the circle elements belong to a single cluster and class) results in an Error rate of 0.00, the *p*-value of the *F*-score is a relatively high 0.341.

Figure 3.9 illustrates the first consideration of the data, wherein each of the three element types (diamond, square, and circle) are assigned to a single cluster and a single class. This arrangement results in no classification errors (so, Error rate = 0.000); however, the relatively large within cluster variability compared to the variance between the means (centroids) of the clusters results in a relatively high *p*-value for the *F*-score, equal to 0.341. The sum of Error rate and $p(F\text{-score})$, then (Equation (3.7)), is 0.341.

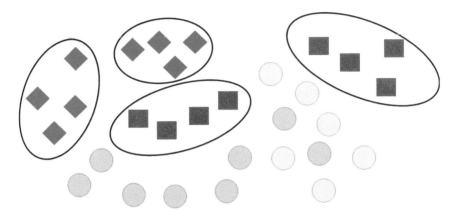

Figure 3.10   Set of data belonging to three classes (diamond, square, and circular elements), with the diamond and square element classes assigned to two clusters each. This results in no misclassification errors since no data belonging to the two classes are outside the boundaries of the clusters. This arrangement (again assuming all of the circle elements belong to a single cluster and class) again results in an Error rate of 0.00, and the p-value of the F-score is a relatively low 0.117.

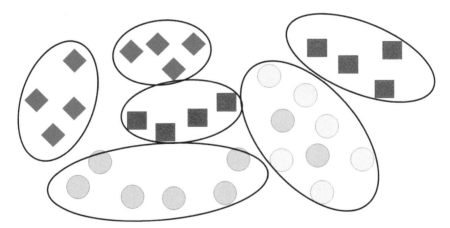

Figure 3.11   Set of data belonging to three classes (diamond, square, and circular elements), with the diamond, square, and circular element classes assigned to two clusters each. In the case of the circular elements, the two clusters are two separate classes. This new arrangement results in two misclassification errors (two of the lighter-shaded circles), and, thus, the Error rate = 0.067 (2 out of 30). This arrangement results in a much lower $p(F$-score) of just 0.022 so that the overall objective function of $p(F$-score) + Error_rate = 0.022 + 0.067 = 0.089 (Equation (3.7)). This is a lower total than for the scenarios in Figures 3.9 and 3.10, implying it is a better fit for the data.

Table 3.10 Sample of three clustering + classification scenarios. The minimum $p(F$-score) is obtained for Hypothesis 3. Lower $p(F$-score) implies better separation between the clusters, while lower Error rate implies better accuracy of the grouping. In this particular example, regularization equation (3.7) is used, so no lambda term for changing the relative contribution of error rate and $p(F$-score) is employed (along the lines of Equation (3.8)).

Hypothesis	$p(F$-score)	Error rate	$p(F$-score) + error rate
1 (Figure 3.9)	0.341	0.000	0.341
2 (Figure 3.10)	0.117	0.000	0.117
3 (Figure 3.11)	0.022	0.067	0.089

Figure 3.10 demonstrates the situation in which the diamond and square classes are considered to comprise two clusters. For computing the ANOVA, these clusters can be treated as independent factors. This significantly increases the F-score, with a concomitant reduction in the p-value of the F-score. In fact, it is reduced nearly threefold, at $p(F$-score) = 0.117 for Figure 3.10. Since the classes are therein "allowed" to comprise two or more clusters, the Error rate is still 0.00 for the scenario outlined in Figure 3.10. This solution, based on either Equation (3.7) or Equation (3.8), is superior to the scenarios of Figure 3.9.

The scenario of Figure 3.10, however, does not address the problem of the class that is represented by circular elements. In Figures 3.9 and 3.10, they are considered a single cluster, and, thus, the Error rate is 0.00, but the F-score is lower than when they are considered to also belong to multiple (two) clusters. The case of assigning them to two clusters, which is the equivalent of assigning them to two classes, is shown in Figure 3.11. The simplest splitting of the circular-element class into two classes is shown in Figure 3.11. Two clusters are formed, each representing a separate class, with the right cluster of circular elements containing, in error, two elements that should belong to the left cluster of circular elements.

In Figure 3.11, then, two elements (the lighter two circles in the right cluster of circular elements) are misclassified, resulting in an overall Error rate of 2/30 = 0.067. Because of the tighter clustering, however, the p-value of the F-score is quite low, at 0.022. Thus, the overall objective function (Equation (3.7)) for the scenario of Figure 3.11 is 0.022 + 0.067 = 0.089, which is less than that of the scenarios for Figures 3.9 and 3.10. These results are tabulated in Table 3.10.

In reviewing Table 3.10, it is worth pointing out the following:

(1) *Determining whether there should be three or four classes — should the circles be split into two classes or left as one.* This is the difference between the scenarios of Figure 3.11 (Scenario 3 in Table 3.10) and Figure 3.10 (Scenario 2 in Table 3.10), respectively. If we use the

general equation, Equation (3.8), to guide us, we see that Scenario 2 is preferable so long as the regularization factor, λ, in Equation (3.8), is greater than or equal to 0.414. Thus, for $0.00 < \lambda < 0.414$, Scenario 2 is preferred; for $0.414 < \lambda < 1.0$, Scenario 3 is preferred. In Equation (3.7), $\lambda = 0.5$ and Error rate and $p(F\text{-score})$ are equally weighted, as described above. This seems, *a priori*, the most reasonable value for λ since both Error rate and $p(F\text{-score})$ have the same range of $0.0-1.0$.

(2) *Determining whether the diamond and square classes should be represented as single or double clusters.* As shown in comparing Scenario 1 in Table 3.10 (corresponding to Figure 3.9) and Scenario 2 in Table 3.10, both the diamond and square classes should be represented as two clusters since this reduces the $p(F\text{-score})$ from 0.341 to 0.117, a relative difference which is unaffected by the value of λ in Equation (3.8).

(3) *Determining which combination of classes (3 or 4) and clusters (3−6) to keep.* As noted in (1) above, representing the data in Figure 3.8 as four separate classes (two for the circular elements), with the diamond and square element classes being assigned to two specific clusters, is generally the optimum assignment for the data.

This section outlines an approach to classification, wherein a class may comprise two or more clusters of data. Each of these clusters can be considered a subclass; however, it may be useful to think of each cluster instead as a category, for which all of the normal processes of data mining and knowledge generation can proceed. If the clusters move apart from each other over time − in absolute measures such as term correlation or in relative measures such as relative distance between these categories compared to distance between the overall class and other classes (another F-score) − then the class can be allowed to split into two classes. This "evolution model" for class definitions aligns well with the general approach to continued training that we recommend for any text-based system. Systems pre-adapted to change over time − both as more training data comes in as well as more information successfully moves through the system − are more robust, more adaptable (by definition), and less sensitive to outliers in the training data. In short, such systems are preferred.

3.5 Applications/Examples

In this section, two applications and their corresponding examples are provided to show the principles overviewed in the previous four sections put into action. The applications chosen are improving the results of search using query expansion and using ANOVA together with cross-correlation to provide image classification. We begin with query synonym expansion.

3.5.1 Query-Synonym Expansion

In this experiment, we combined the use of tf*idf (TD*IDF) or (term frequency multiplied by the inverse of the document frequency) a familiar means of ordering documents by search relevance [Spär72], with the concept of query expansion of the searching of a document corpus. This section illustrates the use of a functional measurement – that is, document search matching – as a means of assessing the effectiveness of different approaches to document classification. The classification approach in this section is based on the use of tf*idf to determine the root set of search terms, followed by the addition of synonyms of these search terms to form the final query set. Querying is then simulated by classifying documents from the correct classes against the original tf*idf terms with a range of $0-20$ additional synonym terms. The steps performed in this experiment are given in the following:

(1) We selected a data set of 10 classes of documents, with each class pared down to a set of 100 documents as described elsewhere [Sims19b]. Fifty documents from each class were assigned to a training set, with the other 50 left over for testing. A random number generator (RNG) was used to assign the documents to training or testing.

(2) Using the simplest TF*IDF definition, we simply computed the percent of words in the document belonging to each term for TF; that is, term frequency. Thus, if the document was "the quick brown fox jumps over the lazy dog," then tf(the) = 2/9 = 0.222, and tf(quick) = tf(brown) = $\cdots$ = tf(dog) = 1/9 = 0.111. Thus, the tf description of the terms in a document are simply the percentages of the words in the documents. These sum to 1.0. The inverse document frequency, idf, is computed in the same way as tf, except that the word list used to compute the percentages is the set of all documents belonging to the document set. Suppose, for simplicity of illustration, that there are two documents in the set, with one being "the quick brown fox jumps over the lazy dog" and the other being "the dog eat dog world is quick to end." For the second document, tf(dog) = 0.222, and tf(the) = tf(eat) = $\cdots$ = tf(end) = 0.111. For the combination of these two documents into a corpus, document frequency, df, is given by df(dog) = 3/18 = 0.167. Likewise, df(the) = 0.167, df(quick)=0.111, and df(brown) = df(fox) = $\cdots$ = df(end) = 0.056. From this, it is easy to compute tf*idf for both documents. For "the quick brown fox jumps over the lazy dog," tf*idf(the) = 0.222/0.167 = 1.333; tf*idf(quick) = 0.111/0.111 = 1.0; df(dog) = 0.111/0.167 = 0.667, and tf*df(brown) = tf*df(fox) = tf*df(jumps) = tf*df(over) = tf*df(lazy) = 0.111/0.056 =

2.0. Similarly, for the second document, the reader can readily calculate that tf*idf(the) = 0.667, tf*idf(dog) = 1.333, tf*idf(quick) = 1.0, and tf*idf(eat) = tf*idf(world) = tf*idf(is) = tf*idf(to) = tf*idf(end) = 2.0.

(3) Now that the tf*idf values for all words in all documents are computed, we can compute the similarity of any new document to the two different documents. This is determined by Equation (3.9), which is effectively the dot product of the tf vector of the new document with the tf*idf vector of the existing document:

$$\text{Similarity} = \sum_{i=1}^{n_{\text{words}}} \text{tf} \left(\text{new document word}\,[i]\right)$$

$$\times \, \text{tf} \times \text{idf}(\text{existing document word}[i]). \qquad (3.9)$$

This is illustrated by example. Suppose we want to decide to which of these two short documents another short document, "a dog is quick to eat when it hungers," is most similar. Note that each of the words in this new sentence have a tf = 0.111. The tf vector is thus {0.111, 0.111, 0.111, 0.111, 0.111, 0.111, 0.111, 0.111, 0.111} for {"a," "dog," "eat," "hungers," "is," "it," "quick," "to," "when"}. For the first sentence, "dog" (tf*idf = 0.667) and "quick" (tf*idf = 1.0) are the only non-zero elements in the tf*idf vector for {"a," "dog," "eat," "hungers," "is," "it," "quick," "to," "when"}; that is, this vector is {0.0, 0.667, 0.0, 0.0, 0.0, 0.0, 1.0, 0.0, 0.0}. The dot product (Equation (3.9)) is, therefore, 0.185, from (0.111)(0.667+1.0). For the second sentence, the corresponding tf*idf vector is {0.0, 1.333, 2.0, 0.0, 2.0, 0.0, 1.0, 2.0, 0.0}, as "dog," "eat," "is," "quick," and "to" are words in common. Here, the dot product (Equation (3.9)) is 0.925, from (0.111)(1.333+2.0+2.0+1.0+2.0). Clearly, since 0.925 > 0.185, the sentence "a dog is quick to eat when it hungers" is considerably more similar to "the dog eat dog world is quick to end" than it is to "the quick brown fox jumps over the lazy dog."

(4) Scale this approach to replace "document 1" with "document class 1," "document 2" with "document class 2," and the like for all of the classes. Instead of a single document, then, the tf*idf vectors are based on the tf vector for all of the words in a class of documents, whilst the idf vector is based on all of the words in all of the classes of documents together.

(5) In a search, however, effectiveness is not measured as much by the highest single dot product result. That is the concern of a traditional classification problem, which is effectively covered by point (3) in this list. With search, the goal is to have the most relevant set of results percolate to the top of the list. As such, we have to consider the {tf}·{tf*idf} dot products as a population and select the top dot

product outliers as the positive results for search matching. For a 10-class problem, then, we select outliers based on the z-score that would be expected to provide a residual area under the curve of $1/N$, where N is the number of classes. Since $N = 10$ in this case, the residual area is 0.0833, for which the corresponding z-value, from a z-table, is $z = 1.282$. Any $\{tf\}\cdot\{tf*idf\}$ values with a z-value, computed from Equation (3.10), greater than or equal to 1.282, are listed as positive search results.

$$z = \frac{\left([\{tf\}\cdot\{tf*idf\}]_{\text{document}} - [\mu]_{\{tf\}\cdot\{tf*idf\}}\right)}{[\sigma]_{\{tf\}\cdot\{tf*idf\}}/\sqrt{N}} \quad (3.10)$$

where $[\mu]_{\{tf\}\cdot\{tf*idf\}}$ is the mean value for $\{tf\}\cdot\{tf*idf\}$ among all the training samples compared to all of the classes, $[\sigma]_{\{tf\}\cdot\{tf*idf\}}$ is the standard deviation for $\{tf\}\cdot\{tf*idf\}$ among all the training samples compared to all of the classes, N is in this case 1 since one sample document is compared to each class in sequence, and $[\{tf\}\cdot\{tf*idf\}]_{\text{document}}$ is the $\{tf\}\cdot\{tf*idf\}$ value for a given document when compared to a given class. With this approach, the mean number of documents returned is the percent of the documents belonging to each class; that is, 10%.

(6) Searching is now performed on the test documents (50 of each of 10 classes) by computing their z-values (Equation (3.10)) for each of the 10 classes. Anywhere $z > 1.282$, the threshold value for the specific experiment, the document is assigned to the positive search set for that class. Note that the same document can show up in the search results for more than one class, and, conversely, a document may not show up in the search results of any class.

(7) Next, we perform the augmented searches. We start off with (see (6)) simply using a set of search/query terms, say $\{a,b,c,d,e\}$, which are the highest-rated tf*idf terms for the class of interest. Next, we augment these to include synonyms, e.g., $\{a,b,c,d,e,f,g,h,i,j\}$, using the WordNet NLTK [Word20] and the Wu-Palmer Similarity [Wu94] approach, selecting the synonyms with the highest similarity ratings. We also tested in sequence to see if removing any of the original terms or synonyms improved the classification results. For example, suppose a reduced set, say $\{b,d,e,g,i\}$, provides the best results. This is then the adaptive set of terms used for classification. Since the classification is done using tf*idf based search approaches, it may be surprising that some of the synonyms might remain after removal of some of the original tf*idf terms. However, this is understandable due to the multi-dimensional interaction between terms and classes and

effectively provides an adaptive principal component analysis (PCA). For the specifics of the simple experiment outlined in this section, search queries based on the key tf*idf terms for each of the 10 classes were expanded using 2, 4, 6, 8, 10, 12, 14, 16, 18, or 20 synonyms to augment the class-specific tf*idf search query (which originally consisted of the top 10 tf*idf terms of each class). For example, a search query for the class "Travel" includes the term "Boat" which has synonyms such as "Ship." The similarity [Word20][Wu94]between "boat" and "ship" is 0.91. All synonymic terms for each of the terms in the original search query have the Wu−Palmer similarity calculated, and the terms with the top 2, 4, 6, ..., 20 weights are added to the search query. The tf*idf values of the synonymic terms had already been pre-computed, and so they were simply included in the dot products (Equation (3.9)) used to determine class membership.

With the specifics of the experiment outlined, or any other search-as-classification experiment, it is important to understand the following key terminologies:

(A) Corpus Size: the sum of all positive results (assigned to the search results) and negative results (not assigned to the search results). In our experiment, the corpus size is 1000 documents, of which 500 were assigned to training and 500 to testing, balanced evenly across the 10 classes of documents.

(B) Positives: the documents that are returned to you (matches) as search results. Because of the experimental design implemented, we expect 50 matches to each "query" since that is the size of the test set for each class. The actual value came out slightly above this for most of the searches, likely due to the discrepancy between the actual distribution of {tf*idf} values and that predicted by a Gaussian distribution.

(C) Negatives: the documents not returned to you as search results (non-matches). In this experiment, we expect 450 non-matches to each "query" due to the selection of z-score threshold of 1.282. As noted for the positives in (B.), the actual number was usually between 400 and 450.

(D) True Positives (TP): documents that are returned to you (matches) as search results that are actually from the correct class. The higher the number of TP, the higher the recall, irrespectively of the (defined below) FP and TN since TP + FN = 1.0.

(E) False Positives (FP): documents that are returned to you (matches) as search results that are from a different (incorrect) class. The more FP, the lower the precision of the class.

Table 3.11 The use of synonym-based query expansion to change the precision (p), recall (r), and F-measure (F) which is the harmonic mean of p and r; that is, $2pr/(p + r)$. F-measure peaks (boldface values, farthest right column) for synonymic expansion of 6–12 synonyms; resulting in 16–22 total search terms before pruning (see text for details).

Number of synonyms	Precision p	Recall r	F-measure
0	0.744	0.390	0.512
2	0.741	0.400	0.519
4	0.783	0.470	0.588
6	0.736	0.530	**0.616**
8	0.684	0.540	**0.603**
10	0.648	0.570	**0.606**
12	0.622	0.610	**0.616**
14	0.568	0.630	0.597
16	0.485	0.650	0.556
18	0.366	0.640	0.465
20	0.300	0.620	0.404

(F) True Negatives (TN): documents that are not returned to you (non-matches) as search results that should not have been, as they are from incorrect classes. The higher the TN, the higher the precision of the class since there are less FP.

(G) False Negatives (FN): documents that are not returned to you (non-matches) that are actually from the class being searched. The higher the FN value, the lower the recall, since as FN increases, TP decreases.

(H) Precision (p): this is the percent of (matching) search results that are useful (belong to the correct class). The value of $p = TP/(TP + FP)$.

(I) Recall (r): the percent of all correctly matched documents (documents from the class that is actually being queried) that are returned as part of the search results. The value of $r = TP/(TP + FN)$.

(J) F-measure (F): the harmonic mean of precision and recall, and the recommended metric for optimization of the search. The value of $F = 2pr/(p + r)$.

The latter three values — that is, p, r, and F — are shown in Table 3.11. Clearly, F-measure peaks in the range of 6–12 added synonyms (that is, expanding the 10-word search query to 16–22 terms). Importantly, adding synonyms above 12 does not further increase accuracy. Also, including less than six synonyms results in lower accuracy. When attempting to prune the set of search terms after the addition of the synonyms, typically two or less terms were removed before peak F-measure was observed. The best results obtained for each of the synonymic expansion are shown in Table 3.11.

The results of Table 3.11 are compelling, though certainly not comprehensive (it is, after all, a "mid-size" data, not a "big" data experiment merely meant to show the possibilities here). First off, the highest precision was observed when the synonymic expansion was 40% of the original query size. Second, the highest recall occurred with synonymic expansion of 160% of the original query size. The mean of these two is 100%, which is right in the midst of the range (6−12 synonyms added) that improved the *F*-measure by the most. In this peak range, *F*-measure improved by roughly 20% over not implementing any synonymic expansion, which corresponds to a reduction in error rate by an equally high 20%. The relatively wide peak range indicates that *F*-measure can be substantially improved while still allowing the system to be tuned for high precision (few FP) or high recall (few FN), depending upon which costs the overall system is more highly dependent.

The specifics of how synonymic search will behave for corpora of different sizes and number of classes is unknown since, as noted, this is relatively small, "disposable" data. However, the data shown here indicate that it can be a significant positive influence on document classification accuracy, with the functional application of search accuracy being used to assess this. The peak value, 0.616, is 0.104 higher than the value when using no synonyms, 0.512. This is a 21.3% reduction in error rate.

3.5.2 ANOVA, Cross-Correlation, and Image Classification

In this relatively lengthy subsection, the relationship between cross-correlation, ANOVA, and classification is explored. In connecting these three topics, further insight into how relationships between documents can be used to structure sequential document-related tasks, such as ordering documents for learning (suggesting the "next best" document to read) or organizing documents into a search structure (providing ranked document reading suggestions), is uncovered. We begin the section with a typical systems engineering/process engineering example of employing ANOVA. Next, we connect cross-correlation of different elements in multiple classes with ANOVA. Finally, we show the connection between ANOVA, clustering, and classification. The latter parts rest on regularization approaches that connect statistics and clustering approaches.

ANOVA is a relatively straightforward, yet, elegant way of assessing how well different groupings, or sets, of data can be statistically distinguished from each other. Effectively, ANOVA assigns variability to its different sources and then compares the relative amount of variability in these sources to assess how data, statistically speaking, should be pooled. This is outlined in Table 3.12.

As shown in Table 3.12, which is simply the single-factor ANOVA interpretation table, the main task of ANOVA is to partition the overall

Table 3.12 ANOVA interpretation table. In a simple single-factor ANOVA, shown here, variability is assigned to one of the two sources: the differences between different factors (or groups, treatments, or clusters), which is shorthanded by "B" for between; and the differences within (shorthanded "W") each of these groups. After the sum squared (SS) error between and within the groups is computed, the mean squared (MS) errors are computed by dividing by the degrees of freedom (DF) in each group. Finally, the ratio of the relative MS error between to within groups is computed as the *F*-score. See text for further details.

Source	DF	SS	MS	F
Between (B) Factor, Treatment, Cluster Error	DFB = $k - 1$	SS(between)	MSB = SSB/DFB	F(DFB,DFW) = MSB/MSW
Within (W) Group Error	DFW = $N - k$	SS(within)	MSW = SSW/DFW	–
Total	DFT = $N - 1$	SS(total)	–	–

variability in a set of data to one of two sources: either the differences between (B) different factors (or groups, treatments, or clusters) or to the differences within (W) each of these groups. The sum of squares between the groups, designated SS(B), is given by Equation (3.11), while the sum of squares within the groups, SS(W), is given by Equation (3.12):

$$SS\,(B) = \sum_{j=1}^{n_{\text{factors}}} n_j (\mu_j - \mu_\mu)^2 \qquad (3.11)$$

$$SS\,(W) = \sum_{j=1}^{n_{\text{factors}}} \sum_{i=1}^{n_j} (d_{ij} - \mu_j)^2. \qquad (3.12)$$

Here, n_j = number in jth group (the size of the jth group), μ_j = mean of jth group, μ_μ = mean of all the group means, d_{ij} = data (i) in jth group, $n_{factors}$ = k, and $N = \sum_{j=1}^{n_{\text{factors}}} n_j$. SS(B), thus, indicates the variability between the means of the groups and can be thought of as the inter-cluster variability. SS(W), on the other hand, indicates the variability within the groups and can be thought of as the intra-cluster variability. After the sum squared (SS) error between and within the groups is determined, these sums need to be divided by the appropriate degrees of freedom, resulting in the mean squared (MS) errors. The degrees of freedom (DF) between and within the clusters are designated DFB and DFW, respectively. DFB = $k - 1$; that is, one less than the number of groups. DFW = $N - k$; that is, the total number of samples minus the number of groups. Combined, DFB + DFW = $N - 1$, the total degrees of freedom. Given this, MS(B) and MS(W), the mean squared error between and within groups, respectively, are computed from Equations (3.13)

Table 3.13   Data set for the electrical resistance of a material in a factory over four consecutive days. Eight samples ($n = 8$) are measured each day, and the mean (μ) and standard deviation (σ) are computed for each day. This data is appropriately analyzed with ANOVA and inappropriately with a *t*-test. Please see text for more details.

Day	Number (n)	Mean (μ)	Standard deviation (σ)
1	8	1.0	0.8
2	8	1.2	0.9
3	8	1.5	1.1
4	8	1.9	0.7

and (3.14):

$$MS(B) = SS(B)/DFB = \frac{SS(B)}{k-1} \tag{3.13}$$

$$MS(W) = SS(W)/DFW = \frac{SS(W)}{N-k}. \tag{3.14}$$

Finally, the ratio of the relative MS error between and within groups is computed as the *F*-score. *F*-score = MS(B)/MS(W). On-line and in-book *F*-score lookup tables are plentiful, and one simply needs the *F*-score value and the pair (DFB, DFW) to be able to assess a given statistical confidence that a minimum of one pair of groups have different means.

Let us show the value of ANOVA in an industry setting by example. Suppose that we wish to test the electrical resistance of a material in a factory four days in a row. We notice that the mean resistance is different for each day and so accumulate the data shown in Table 3.13. Note that eight samples are measured each day so that $N = 32$. Since there are four days, $k = 4$. The mean (μ) +/− standard deviation (σ) for those four days are 1.0 +/− 0.8, 1.2 +/− 0.9, 1.5 +/− 1.1, and 1.9 +/− 0.7, respectively. There is a trend for the mean value to be increasing from one day to the next; however, the high standard deviation (coefficient of variance ranges from 0.37 to 0.80 for the four days) means that we must test to see if this trend is statistically relevant.

In order to compute the relevant ANOVA information, Table 3.13 is expanded into Table 3.14. Two extra columns are added: the variance, which is just the square of the standard deviation, and the sum of squared errors within the group (SSW), which from the definition of variance, σ^2, and Equation (3.12), is simply $(n-1)\sigma^2$. That is, SSW for each of the four days is $(n-1)$ times the variance since the variance is simply the sum of squared errors divided by $(n-1)$. With ANOVA, we test whether or not there is a statistically significant difference between at least two of the groups. Our null hypothesis, H_0, is that there is no statistically significant difference, and our alternate hypothesis, H_a, is that there is a statistically significant difference between at least two groups.

Table 3.14 Data set of Table 3.13 with two columns added: one for variance and one for the sum of squared errors within each day's sample set (SSW).

Day	Number (n)	Mean (μ)	Standard deviation (σ)	Variance (σ^2)	SSW $= (n - 1)\sigma^2$
1	8	1.0	0.8	0.64	7(0.64) = 4.48
2	8	1.2	0.9	0.81	7(0.81) = 5.67
3	8	1.5	1.1	1.21	7(1.21) = 8.47
4	8	1.9	0.7	0.49	7(0.49) = 3.43

Table 3.15 ANOVA interpretation table for the data in Table 3.13. Please see text for details.

Source	DF	SS	MS	F
Between (B)	DFB $= k - 1 = 3$	SS(B) = 3.68	MSB = 1.2267	$F(3,28)$ = MSB/ MSW = 1.558
Within (W)	DFW $= N - k =$ 28	SS(W) = 22.05	MSW = 0.7875	–
Total	DFT $= N - 1 =$ 31	SS(T) = 25.73	–	–

Also from Table 3.14, μ_μ is readily obtained from (1.0 + 1.2 + 1.5 + 1.9)/4 = 1.4, from the column of means (μ). We next compute the main measurements for the F-score. SS(W), from Equation (3.12), is simply the sum of the sixth column in Table 3.14, and, therefore, 22.05. SS(B), from Equation (3.11), is simply $8((1.0 - 1.4)^2 + (1.2 - 1.4)^2 + (1.5 - 1.4)^2 + (1.9 - 1.4)^2$), using μ_μ and the column of means (third column), which is 8(0.16 + 0.04 + 0.01 + 0.25) = 3.68. Next, MS(W) = SS(W)/DFW, where DFW = 32 − 4 = 28, which is 7 degrees of freedom for each of the eight-element groups, as should be. MS(W) = 22.05/28 = 0.7875. MS(B), similarly, is SS(B)/DFB, where DFB $= k − 1 = 3$, one less than the number of groups. MS(W) therefore = 3.68/3 = 1.2267. Finally, the F-score with df = (3,28) is MSB/MSW = 1.558. These data are cataloged in Table 3.15.

From a table of F-scores, we have the following values:

(1) for $p < 0.100$, $F > 2.29$;
(2) for $p < 0.050$, $F > 2.95$;
(3) for $p < 0.025$, $F > 3.63$;
(4) for $p < 0.010$, $F > 4.57$;
(5) for $p < 0.001$, $F > 7.19$.

It is clear from Table 3.15 that there is no statistical support from the F-score for there being a statistically significant difference between any two groups (including that samples from day 1 compared to the samples from day 4) since $p \gg 0.100$ for $F(3,28) = 1.558$. In fact, the p-value for $F(3,28)$ is 0.222. The result is quite clearly not significant at $p < 0.20$, let alone <0.05.

Before continuing with cross-correlation, it is worth noting here in a world that can no longer simply accept as a conclusion "$p < 0.05$" [Wass19] how statistics could have been misused in this example. This data was taken from an actual quality assurance procedure used to analyze the electrical resistance of a conductive ink by one of our colleagues. That colleague was concerned that the mean resistance of the ink was rising each day over the four days. In particular, this colleague compared the values for day 1 (1.0 +/− 0.8) with the values for day 4 (1.9 +/− 0.7) using a *t*-test (Equation (3.15)).

$$t = \frac{|\mu_2 - \mu_1|}{\sqrt{\frac{1}{n_1} + \frac{1}{n_2}} \sqrt{\frac{(n_1-1)\sigma_1{}^2+(n_2-1)\sigma_2{}^2}{n_1+n_2-2}}}. \tag{3.15}$$

Using Equation (3.15), if we compared day 1 to day 4, $t = 2.395$, with df = 14, for which $p = 0.0312$ (two-tailed). This appears to provide a $p < 0.05$ statistical significance, but it disregards the fact that days 2 and 3 have higher variability than days 1 and 4, along with the fact that the experimental design included all four days, and so must account for them all in the statistical analysis. Both of these require ANOVA and not selective application of *t*-tests. The *t*-test results, however, were reported as fact in a meeting attended by one of the authors. In isolation, the *t*-test makes sense. In context, none at all. This is why the expression, "There are three kinds of lies: lies, damned lies, and statistics," often erroneously attributed to Disraeli, is so relevant to this day. In today's data-driven society, however, statistics can no longer be used to deceive. Please use ANOVA whenever possible in place of a *t*-test or *z*-test.

Back to our main thread. We would now like to apply some of what we have learned toward cross-correlation and the arts of clustering and classification. A simple experiment, involving the normalized cross-correlation values computed in comparing seven elements from two classes of data, is presented to illustrate the approach functionally (our preferred method in this text!). Table 3.16 comprises the normalized cross-correlation values and, as is required for such a set of calculations, one notes each element $C_{ij} = C_{ji}$ in the matrix. Also, the diagonal elements by necessity are equal to 1.0 since a data element has 100% correlation with itself. The data shown in Table 3.16 are taken from seven equally sized images (all have the same *N*, which is equal to the height in pixels multiplied by the width in pixels). A similar set of data could readily have been obtained from seven documents, except that Equation (3.17) would have been applied to the word percentage histograms of documents *x* and *y*.

$$\text{Norm\_Corr}(x, y) = \frac{\sum_{n=0}^{N-1} x[n] \times y[n]}{\sqrt{\sum_{n=0}^{N-1} x[n]^2 \times \sum_{n=0}^{N-1} y[n]^2}} \tag{3.16}$$

Table 3.16 Cross-correlation data for a set of seven samples. The data shown was originally obtained from a set of seven images from two classes. The data presented here is the normalized cross-correlation between any pair of two images, as defined by Equation (3.16). However, for text analytics, a similar matrix could have been computed as normalized word correlation between documents (Equation (3.17)).

	1	2	3	4	5	6	7
1	1.0	0.85	0.43	0.56	0.83	0.33	0.84
2	0.85	1.0	0.29	0.44	0.91	0.37	0.95
3	0.43	0.29	1.0	0.88	0.36	0.93	0.31
4	0.56	0.44	0.88	1.0	0.49	0.97	0.40
5	0.83	0.91	0.36	0.49	1.0	0.46	0.96
6	0.33	0.37	0.93	0.97	0.46	1.0	0.17
7	0.84	0.95	0.31	0.40	0.96	0.17	1.0

$$\text{Word\_Corr}(x, y) = \frac{\sum_{n=0}^{N-1} p_x[n] \times p_y[n]}{\sqrt{\sum_{n=0}^{N-1} p_x[n]^2 \times \sum_{n=0}^{N-1} p_y[n]^2}}. \qquad (3.17)$$

In Equation (3.16), N is the size of the image in pixels, x and y are the two images, and $x[n]$ and $y[n]$ are the appropriate (e.g., intensity) values of pixel n in each of the two images, respectively. The sum shown, Norm_Corr(x,y), is the normalized cross-correlation of the two images and has a range of values [0,1].

In Equation (3.17), $p_x[n]$ and $p_y[n]$ are the percentage of word n out of the total number of words in each document x and y, respectively. N is the total set of unique words in the combined set of two documents, and Word_Corr(x,y) is the normalized cross-correlation of the two word sets (or "documents"), also having a range of values [0,1]. When two documents have no words in common, Word_Corr$(x,y) = 0.0$, and when two documents are identical in their word histograms (usually only the case when they are identical documents), Word_Corr$(x,y) = 1.0$.

Once the table of cross-correlation data is obtained, either image or document classification proceeds along the same lines. As an overview, we are interested in performing an ANOVA on the cross-correlation matrix. Glancing at the matrix in Table 3.16, it is clear to even a casual observer that samples {1,2,5,7} form one cluster and samples {3,4,6} another. In general, determining the "optimal" assignment to clusters can be handled along the lines of the approach given earlier (Equation (3.8) and the text around it). For this example, we simply assert that out of all possible combinations of two or three clusters, constrained only by the requirement that each has a minimum of two samples, the two-cluster representation of {1,2,5,7} and {3,4,6} has

Table 3.17   Cross-correlation values for all pairs of elements within the same cluster for the data of Table 3.16, and the corresponding distance values, computed from Equation (3.18).

Within cluster distances	
Cross-correlation	**Distance**
0.85	0.15
0.83	0.17
0.84	0.16
0.91	0.09
0.95	0.05
0.96	0.04
0.88	0.12
0.93	0.07
0.97	0.03

the highest $p(F$-score). This interpretation, therefore, represents the optimal clustering for the data in Table 3.16. Given this, we turn to the actual data in the table and note that the table does not contain data representative of any representation of "distances" between the samples. As a consequence, we have to somehow use the cross-correlation data to generate data usable as distance information. We note that the values within the clusters – that is, $\{0.85, 0.83, 0.84, 0.91, 0.95, 0.96\}$ and $\{0.88, 0.93,$ and $0.97\}$ in comparing amongst $\{1,2,5,7\}$ and $\{3,4,6\}$, respectively – are quite different from the values between the clusters: $\{0.43, 0.56, 0.33, 0.29, 0.44, 0.37, 0.36, 0.49,$ $0.46, 0.31, 0.40, 0.17\}$. It is clear that some form of inverse operation is required to transform these cross-correlation values into distance. Since the cross-correlation formulas (Equations (3.16) and (3.17)) are the ratio of two second-order functions, they provide as output a normalized distance between 0.0 and 1.0. Thus, we have a relatively straightforward operation to remap the cross-correlation to distance, given by Equation (3.18):

$$\text{Distance} = 1.0 - \text{Corr}(x, y). \tag{3.18}$$

Here, Corr(x,y) is either Norm_Corr(x,y) as in Equation (3.16) or Word_Corr(x,y) as in Equation (3.17). At this point, the source of the Corr(x,y) information no longer matters, and we proceed identically with either image or text analytics. Table 3.17 collects the distances within the clusters, and Table 3.18 collects the distances between the clusters.

In a situation like this example, since we are comparing two populations with what appear to be Gaussian distributions, a t-test (Equation (3.15)) is a simple means of assessing the difference in the distance populations. For the within-cluster distances, mean $\mu = 0.098$, standard deviation $\sigma = 0.051$, and $n = 9$. For the between-cluster distances, mean $\mu = 0.616$, standard deviation $\sigma = 0.099$, and $n = 12$. However, a t-test requires degrees of freedom to be

Table 3.18   Cross-correlation values for all pairs of elements in two different clusters for the data of Table 3.16, and the corresponding distance values, computed from Equation (3.18).

Between cluster distances	
Cross-correlation	**Distance**
0.43	0.57
0.56	0.44
0.33	0.67
0.29	0.71
0.44	0.56
0.37	0.63
0.36	0.64
0.49	0.51
0.46	0.54
0.31	0.69
0.40	0.60
0.17	0.83

2 less than the total number of samples; so we use $n_1 = n_2 = 3.5$, or half of the samples, to be able to use the t-statistic as an approximation to an actual t-test. For the data given in Tables 3.17 and 3.18, $t = 8.702$, df $= 5$, for which $p = 0.000332$. This is the lowest observed p-value for any comparison of 2 or 3 clusters of 2+ elements from the set of seven elements in Table 3.16. This is a first approximation to finding an optimal cluster formation that incorporates cross-correlation, but we can do better.

Next we convert the table of cross-correlation values into an ANOVA, using distance values that have to satisfy the following criteria:

(1) The distances must be representative of the proximity of the elements to one another.
(2) The degrees of freedom must match the degrees of freedom of the data.
(3) The different potential outputs must be meaningfully and consistently comparable relative to the other outputs.

Because of the nature of the ANOVA, we will automatically generate a statistical probability, p, for the test data, which helps us satisfy criterion (3) without special effort. Table 3.19 lists some of the minimum F-scores for df $= (k - 1, N - k) = (1,5)$ required for various levels of statistical confidence ranging from $p = 0.001$ to $p = 0.100$.

In order to ensure that (1) the distances are representative of the proximity of the elements to one another, we produce a table of mean distances to elements in the cluster and mean distances to elements in the nearest cluster. The latter is just the "other" cluster in this case since there are only two clusters, but, in general, the nearest other cluster should be used. These values are catalogued in Table 3.20.

Table 3.19 Minimum $F(1,5)$ values required to achieve various levels of statistical confidence for the two-cluster, seven-element example originally in Table 3.16.

p-Value	Minimum $F(1,5)$
0.100	4.06
0.050	6.61
0.025	10.01
0.010	16.26
0.001	47.18

Table 3.20 Mean distance to other elements in the same cluster and to elements in the nearest other cluster for the data in Table 3.16. Since there are only two clusters in this example, the other cluster is all the elements not in the same cluster.

Element	Mean distance to other elements in the same cluster	Mean distance to elements in the nearest other cluster
1	0.160	0.560
2	0.097	0.633
3	0.095	0.653
4	0.075	0.528
5	0.100	0.563
6	0.050	0.668
7	0.083	0.707

In order to ensure that (2) the degrees of freedom match the degrees of freedom of the data, we provide the mean distances as in Table 3.20 so that there is only one distance per sample and then use Equations (3.13) (for between cluster) and (3.14) (for within cluster) sums of squared error. The SS(between) is simply calculated from $(0.653 + 0.528 + 0.668)/3 = (0.560 + 0.633 + 0.563 + 0.707)/4 = 0.616$, which is the mean distance to other samples not in the same cluster. This distance is used for the "squared error," and so taking into account the nj values in each cluster j, SS(between) $= 7(0.616) = 4.312$, which is also just the sum of the rightmost column ("Mean distance to elements in the nearest other cluster") in Table 3.20. Mean squared error between, or MS(between), is readily calculated from SS(between)/DFB = SS(between)/$(k - 1) = 4.312/(2 - 1) = 4.312$. Next, SS(within) is calculated for the data. Again, this is simply a column – in this case, the middle column or "Mean distance to other elements in the same cluster" – sum in Table 3.20. This sum, SS(within), is 0.660. MS(within) is simply SS(within)/DFW = SS(within)/$(N - k) = 0.660/(7 - 2) = 0.132$. Finally, the F-score is computed. $F(1,5)$ = MSB/MSE $= 4.312/0.132 = 32.67$. For this F-score, $p = 0.00229$, which is a lot more conservative than the t-test (where $p = 0.000332$), although the variability between the two clusters is highly significant statistically either way.

Why is the F-score so much more conservative than the *t*-test? As an indirect answer, we note that the *F*-test could also have been performed using distance squared as the SSE. With that approach, it is left to the reader as an exercise to find that for the within-cluster distances-squared, mean $\mu = 0.0122$, standard deviation $\sigma = 0.0104$, and for the between-cluster distances-squared, mean $\mu = 0.3890$, standard deviation $\sigma = 0.1257$, for which $t = 5.055$, df = 5, $p = 0.0039$. For the distances-squared, the SS(between) = 7(0.389) = 2.723. MS(between) = SS(between)/DFB = SS(between)/($k - 1$) = 2.723. SS(within) = 0.0796, and MS(within) = SS(within)/DFW = SS(within)/($N - k$) = 0.0796/(7 $-$ 2) = 0.0159. From these, the *F*-score = MSB/MSE = 2.723/0.0159 = 171.3. For this *F*-score, $p = 0.000046$, which is a lot **less** conservative than the *t*-test (where $p = 0.0039$). Thus, it is clear that these approaches are only approximations to actual *t*-tests and ANOVA since they do not necessarily track together. But, they certainly give us a lot of flexibility for the formation of clusters and classes. This flexibility, as it relates to the definition of distance in the process outlined in this section, is encapsulated by Equation (3.19):

$$\text{Distance} = (1.0 - \text{Corr}(x, y))^\alpha. \tag{3.19}$$

3.6 Test and Configuration

The previous section opens up a set of possibilities for test and configuration where text analytics benefit from statistics, machine learning, or both approaches. If the distance between elements are to be considered as one aspect of the overall clustering and classification approach, then system optimization (on training data) may consist of finding the right power to raise the Cartesian distance to provide an optimal behavior (such as accuracy on labeled data, equal *p*-values for the *t*-test and *F*-score, etc.).

Other configurations of text data that can take advantage of machine learning is the possibility of using document content to create images synthetically and then use convolutional neural networks or other successful deep learning approaches for image recognition to perform text recognition. Part of text analytics may also be based on "neural carving," which is part of the normal development process in brains. With the synonymic expansion, we observed that cutting back from the combined set of search terms and their expanded set of terms sometimes provided better results. Regardless, the results shown in this chapter indicate that there are still a lot of possibilities unexplored in text analytics.

3.7 Summary

This chapter describes some existing and some novel methods for text clustering, classification, and categorization. Regularization, necessary to prevent the explosion of coefficient values in many machine learning procedures, was shown valuable both in model building approaches such as regression as well as clustering. The tradeoff between error rate and cohesiveness of clusters was shown to be an application for simultaneously optimizing clustering and classification. The chapter finished with two more in-depth examples: the first addressed synonymic expansion as a means to optimize text corpus behavior, and the second explored the relationship between ANOVA, cross-correlation, and clustering. At the chapter's end, the reader should have several more tools for optimizing the structure of a large text corpus. The regularization and statistical approaches introduced here, moreover, may have general value in a wide range of machine learning applications.

References

[Demp77]   Dempster AP, Laird NM, Rubin DB, "Maximum Likelihood from Incomplete Data via the EM Algorithm," *Journal of the Royal Statistical Society*, Series B, **39**(1):1–38, 1977.

[Sims13]   Simske S, "Meta-Algorithmics: Patterns for Robust, Low-Cost, High-Quality Systems," Singapore, IEEE Press and Wiley, 2013.

[Sims17]   Steven S, Vans M, "Learning before Learning: Reversing Validation and Training," ACM DocEng 2017: 137-140, 2017.

[Sims19a] Simske S, "Meta-Analytics: Consensus Approaches and System Patterns for Data Analysis," Elsevier, Morgan Kaufmann, Burlington, MA, 2019.

[Sims19b] Simske SJ, Vans M, "Functional Applications of Text Analytics Systems," *Archiving 2019*, 116-119, 2019.

[Spär72]   Spärck Jones K, "A Statistical Interpretation of Term Specificity and Its Application in Retrieval," *Journal of Documentation* **28**: 11–21, 1972.

[USGS20]   U.S. Geological Survey, "The National Map," https://viewer.nationalmap.gov/advanced-viewer/, accessed 16 April 2020.

[Wass19]   Wasserstein RL, Schirm AL, Lazar NA, "Moving to a World Beyond 'p < 0.05'," *The American Statistician*, 73(sup1):1-19, 2019.

[Word20]   WordNet   NLTK,   https://www.nltk.org/howto/wordnet.html,
            accessed 23 March 2020.
[Wu94]   Wu Z, Palmer M, "Verb semantics and lexical selection,"
            *Proceedings of the 32nd Annual Meeting of the Associations for
            Computational Linguistics*, pp 133-138, 1994.

4

Translation

> "*Woe to the makers of literal translations, who by rendering every word weaken the meaning! It is indeed by so doing that we can say the letter kills and the spirit gives life*"
> – Voltaire

> "*Humor is the first gift to perish in a foreign language*"
> – Virginia Woolf

Abstract

Translating from one dialect to another is tricky enough; translating from one related language to another is fraught with error; and translating between two non-related languages may not even be possible. In this chapter, we consider some of the problems with translation, along with some of the possibilities for improving upon existing methods. We then address functional means of assessing the efficacy and accuracy of translation, with various amounts of human intervention in the process. We end up introducing the concept of "acceptable substitution" as a functional replacement for direct translation.

4.1 Introduction

Without translation, we effectively have two choices for our communications: (1) everybody learns to speak the same language (aka linguistic homogenization) or (2) people are segregated by their languages. Neither of these provides a satisfactory solution to linguistic diversity. For one thing, no language – not even English or Mandarin – offers a sufficiently large percentage of the earth's population to mandate for an exclusion of the rest of the languages. Second, the diversity of languages on the planet also provide for tremendous diversity of thought. Translation, in trying to map the thoughts and sequence of thoughts from one mind to other, can be used to provide insight into both the past and the future of thought.

Plato, Laozi, Avicenna, and Marie Curie shared no languages in common, yet all of us benefit from their combined wisdom. Translation is not a simple mechanical process; it is not a mapping of terms from one syntax to another. Instead, translation is the way in which systems of thought can be made mutually understandable. Translation is, in short, linguistic empathy.

Linguistic empathy may not be what Voltaire said, but he would certainly concur that literal translations have none. Linguistic differences arise along almost any continuum that has a range of values, including gender, age, race, class, and location within a single language's range. Think of teenagers and their parents, unintentionally and often intentionally speaking two different dialects of the same language in the same home. Slang and jargon are not always born of necessity; they are also a (sometimes not so) subtle means of empowering a group which does not feel appreciated, equitably treated, or understood in some context. An adolescent may be particularly drawn to neologisms and de novo idiomatic expressions particularly when they feel that the older generation does not understand them anyway.

Regardless of the sociological reasons for linguistic diversity, it seems feckless for some form of "language police" to dictate linguistic rules, let alone uniformity, within a specific language. French-speaking Canadians do not speak French; they speak Quebecois. Natives of the Rio Grande valley may speak Spanglish rather than polarizing their verbiage to Spanish or English. To paraphrase Voltaire, the more languages change, the more they stay the same. Change in language allows each successive generation to leave its mark on the culture and to recreate humor from the ground up. The younger generation cringes at the previous generation's humor, calling it (tellingly) "dad jokes," in part because the humor of the previous generation is so integral to the vernacular of that same antecedent lingo. Virginia Woolf noted that humor is the first part of language lost in translation, and this is almost by definition since so much of humor relates to the actual usage of language: puns, plays on words, acronyms, and malapropisms are, superficially, no more difficult to translate than any other part of language; however, because they relate the meaning of two or more idioms somewhat randomly connected, the odds are low that a similar relationship exists in the language translated into. A simple example helps illustrate that. A person with a 30-handicap might comment after a particularly bunker- and wood-filled round on the links that "golf is flog spelled backwards," which certainly does not work in German, in which "golf" is still "golf" but "flog" is "prügeln" instead.

Translation of known languages, therefore, is a difficult science. Now, imagine trying to translate a language about which very little is known. People a lot smarter than us (and even if you do not believe that, they certainly have a lot more specialized linguistic skills than we do) still have not been able to decipher Linear A and B, even when they presumably know much about the lineage of these two languages. In order to appreciate the difficulty of translation, we must first describe what is in fact known about them. Linear A is the name given to the script used by the Minoans (inhabitants of the island of Crete) from the time period running from approximately 1800 to 1450 BCE. It was applied to palace and religious objects and writings and is assumed to therefore represent the language of the Minoans. This is an assumption, of course, as it may have been, like some earlier scripts of Mesopotamia, used for very prosaic tasks that are only loosely associated with language. One of these, the Sumerian cuneiform, which dates back 10,000 years, was apparently originally derived from cataloguing of trade goods and livestock on clay tablets. This is an interpretation that makes a lot of sense since when your medium is clay, you probably are not planning to sit down and write "The Brothers Karamazov." It needs to be something with high value per pictograph, and warehousing certainly fits that bill. The Cretan script we know as Linear A was discovered by Sir Arthur Evans sometime around 1900. Further exploration of Cretan antiquities led to the identification of Linear B, which has been translated and shown to represent Mycenaean Greek. This was the Bronze Age Greek that was used from roughly 1600 to 1100 BCE. Even with the Linear B translation, however, no texts in Linear A have been deciphered.

Linear A may remain forever untranslated or at least never be more than speculatively translated. This stems in part from the fact that Linear A grew as a language in a manner independently from that of the Egyptian and Mesopotamian language systems. According to Younger [Youn00], there were four major branches of proto-Greek in the second millennium BCE. These have been called Linear A, Linear B, Cypro-Minoan, and Cretan, and each of them is represented by hieroglyphic pictographs. Hieroglyph means "sacred writing" and is an acknowledgement of the fact that literacy was a precious commodity in ancient times. In the 1950s, when Linear B was translated as Mycenaean Greek, the translators noticed the many shared symbols with Linear A. If they notate similar syllabic values, this has not helped since Linear A still has not been successfully translated. The key to this tale of Linear A and B is that some ancient languages can be translated because they have one or more of the following conditions that are met:

(1) Sufficient samples of the language exist;
(2) Related languages exist and/or have been translated;
(3) Context for the writing can be/is established;

Generally, if condition (1) is met, adoption of the language was widespread enough that both (2) and (3) exist. In other words, the limiting factor for the promise of successful translation is often (2) and sometimes (3). The famed Rosetta Stone, now eponymous for learning a second language, as one example provided both (2) and (3). For (2), it provided the same information in Ancient Egyptian (hieroglyphics) and Demotic (Egyptian between the hieroglyphics and Coptic era of writing systems), along with Greek. This allowed ready translation of the ancient hieroglyphics and effectively "exposed" the secrets of the pyramids. Context (3) was also provided inasmuch as the three parts of the Rosetta Stone contain the same information, thereby providing triple alignment between the three possible pairings. Additional context is also provided from the purpose of creating the stone in the first place: to celebrate the first anniversary of the coronation of the Egyptian pharaoh Ptolemy V in 196 BCE.

Languages that are lost, therefore, share three factors in common: (1) fragmentary samples of them remain; (2) they are unrelated to languages for which a large sample set is available (which includes all modern languages, of course); and (3) the context around the writings may not be understood. Given the massive growth of supercomputing, artificial intelligence, and deep learning over the past few decades, linguists have far more powerful tools to analyze languages than at any time in history. However, with these three factors in place, trying to "reverse engineer" a dead language (one that is no longer spoken by anyone living) is analogous to trying to decrypt a sample of encrypted text without knowing the private key of the encryption algorithm. With a fully entropic (randomizing) encryption algorithm, all possible "translations" (that is, decryptions) are equally likely, and so no relevant translation can be made.

The value of context can be illustrated by the "Stranger Things" science fiction horror web television series, which first aired in 2016. This show is a "retro" show, set in the 1980s, or more than 30 years before the actual filming. Because the Duffer brothers and the rest of the writers of the show could not, apparently, imagine a world in which instant communication at a distance was unavailable to its characters (a very real condition of the 1980s), the show uses unrealistic long-range, always-on, walkie talkies to allow the characters to communicate like the youth of the 2010s are accustomed to. Of course, it is a science fiction/horror show, so it is not supposed to be realistic, but aside from the supernatural elements, the show is supposed to represent life in the 1980s. The walkie talkies in the show are not contextual to the 1980s. If it is this difficult to represent a world only 30 years ago, imagine the difficulty of representing a world 3000 or more years ago, and in so doing trying to understand the nuances of their language. Surely, any interpretation of a language as ancient as Linear A must fit the facts tightly before we can accept the translation as valid. Thus, the lack of additional facts – text samples, translations, and/or context – prevents a dead language from being resurrected.

Fortunately, in the modern world, it is a lot easier to translate between two languages that are both spoken, right? Well, that depends on the competency of the translator. As an example of translation difficulties, let us here consider the "Jimmy Carter translator story" from 1977 [Gues14]. In what we will take for veracity here, the reference notes that "On a state trip to Poland in 1977, US President Jimmy Carter explained how he wanted to learn more about the Polish people's desires for the future, both politically and economically." How hard can that be to translate? Well, apparently, a little harder than one might expect. Carter hired a Polish translator, but apparently not a polished translator. The above quote was translated to imply that Carter was not interested in just any of the Polish people's desires, but specifically in their carnal desires. That was strike one. Strike two was when the translator, attempting to tell the audience that Carter was returning to America, instead told them that Carter was abandoning America. Carter stayed at the plate, though, but strike three came soon after, when the translator used numerous Russian words in place of Polish words. On the bright side, he may have gotten those words right in Russian! Regardless, this story shows that even a correct translation can be the wrong thing in certain context.

Combined, these anecdotes show that translation is not a simple algorithm. Translation, instead, requires context and, as we saw for the Jimmy Carter in Central Europe fiasco, it requires a failsafe. Thus, translation is a systems-level approach, requiring an integrated test and measurement, reliability, and robustness plan. Test and measurement, where possible, can be tied to functional measurements. A simple functional measurement is to gauge the response for agreement with the intended translation. One can imagine President Carter's reaction when Polish men appeared angry over the use of Russian words, and the Polish women appeared angry (or worse, interested!) when his need for knowledge became carnal. Clearly, failsafes to at least indicate when translation has gone horribly awry are needed.

In order to address these types of translation failures, we propose a system for translation consisting of the elements outlined in Table 4.1. There are five main elements, which are described in the following paragraphs. For each of the elements, "out of bounds" or anomalous behavior from the behavior expected for the semantics/emotional content of the original text are computed and scored. Each of the five factors in Table 4.1 may be scored, for example, from 0 to 10, then multiplied by the relative weight of each of the factors in a training set, and then an overall score for anomalous behavior obtained from the sum of these factors. Any other suitable objective function based on these five variables may, of course, be used instead.

Before employing these five factors, the expected emotional response of the text is computed using textual emotion detection. Recently, the identification of text emotion has been explored by deep learning [Wang20], and it has been a subject of intensive research for well over a decade [Wu06]. Although hardly perfect, emotion detection accuracies for text, face, and speech continue to get better over time, and accuracies above 90%, which is meaningful for a failsafe system, are consistently reported. The expected emotion of the original text can be reinforced by determining the emotion on the face and in the voice of the speaker such that we are fairly confident on the identified emotional content of the pre-translated text. Additionally, the emotional content of the original-language content can be specified using Emotion Markup Language, a W3C Recommendation since 2014 [Emot14].

Next, the intended emotional effect of the original language is compared to the actual emotional response using the five factors in Table 4.1. The first factor in this "failsafe" system is face emotion recognition. Algorithms for facial emotion detection have been in existence for several decades [DeSi97] and have continued to improve [Nagp10]. The face emotion recognized is compared directly to the expected emotion of the original text. The degree

Table 4.1 Elements of a functional translation system and their roles in the system. Please see text for details.

Element	Role of the element in the functional translation
(1) Face emotion recognition	Detected emotion from the facial images is compared to the emotions predicted in response to the textual content of the original language
(2) Voice emotion recognition	Detected emotion from the vocal analysis is compared to the motions predicted in response to the textual content of the original languages
(3) Textual cohesiveness	Each translated section of text is compared to the text around it to test for consistency of content
(4) Response interval in speech	Intervals during speech and between one speaker and the other are analyzed to determine translator difficulties and anomalous processing of the information by the recipients of the translation
(5) Alternative translation elimination	Alternative (incorrect) translations are gauged for their expected response: if an alternative is confirmed to be the best match for the actual set of Elements 1−4 above, then a prepared apology and replacement translation are delivered

of discrepancy can be assessed. For example, if the original-language text is meant to convey happiness, and the facial emotion of the recipient of the translated text is happy, then the difference (or *penalty*) score is 0. If the recipient is unemotional (or blasé), then the penalty score might be 2; if the recipient is confused, the penalty score might be 5; and if the recipient is unhappy, angry, or disgusted, the penalty score might be 10. The penalty score is then multiplied by the relative weight of the "face emotion recognition" score which may be proportional, for example, to the inverse of the error rate. The error rate is simply 1.0 − accuracy. This score is the **confidence score** for the face emotion recognition.

The second factor in the failsafe system is the voice emotion recognition. This technology has been around for a few decades and was worked on by one of the authors [Yaco03]. Accuracy is routinely above 90%. In this early work [Yaco03], the particular focus of the voice recognition was on extracting the emotional features from short utterances typical of interactive voice response applications, making it immediately applicable to a sentence-by-sentence update preferred for real-time translation along the lines of President Carter's unfortunate adventure. Again, the penalty score between the emotion of the original-language and the translation is gauged, and given a score from 0 to 10. This score is then multiplied by the confidence score of the voice emotion recognition in the same manner as for the face emotion recognition.

The third factor is the measurement of text (emotional) cohesiveness. This is an approach that usually results in lower accuracy than simple emotional assessment since the cohesiveness depends on at least two (usually consecutive) text phrases matching each other for emotion or content. If the accuracy of each text phrase is 90% and each assessment is independent, then the agreement for two consecutive phrases is 82%, from $0.9 \times 0.9 + 0.1 \times 0.1$ (assuming all of the inaccurate results are identical). Penalty scores are determined as for the other factors. As a consequence of this predicted (though still experimentally verified) lower accuracy, the confidence score for this factor will generally be lower than that for the previous two factors.

The fourth factor to be considered in the failsafe system is the response interval in speech between the translator and the recipient of the translator's speech. When there is an inordinate pause between the recipient hearing the text and replying to the text, this is a sign that something is wrong with the conversation. Thus, in this step, the conversational intervals are analyzed to determine translator difficulties on the original-language side and the difficulty in understanding the translated information by the recipient. The penalty score can be assigned a value of 0 to 10 based on how many standard deviations away from the expected interval the delay actually is.

For example, a delay of expected interval is assigned 0, and a delay of three or more standard deviations longer than predicted is assigned a score of 10. In between, intervals of one and two standard deviations longer than predicted are assigned penalty scores of 3.3 and 6.7, respectively. The penalty scores, as usual, are then multiplied by the confidence score for this factor. The confidence score for this factor will often be lower than those of the previous three factors due to the many sources of variability incorporated into the intervals.

The fifth and final factor involved in the failsafe system is the elimination of alternative translations. In this step, candidate alternative translation phrases are created and measured for their emotional content. If any of the candidate translations is deemed a better match for the emotional content of the original-language text, then a penalty score is assigned to the original translation. The penalty score can be assigned either open-loop or closed-loop. In the open-loop system, the better emotional matching to the original-language text is given a penalty score based on the emotional distance as described above; e.g., 0 for a matching emotion; 2 for a slightly different emotion; and 5 or more for a greatly different emotion. For the closed-loop system, the new translation can be attempted with an introduction such as a prepared apology, "Sorry, I meant..." Then the alternative translation can be assessed for the first four factors as described above and relatively compared to the original translation.

This example simply illustrates how translation system can be made more robust by placing them in context of failsafe approaches and by allowing the possibility of automatic apology and a second attempt at translation. With this introduction, we now consider the current state of automatic translation.

4.2 General Considerations

4.2.1 Review of Relevant Prior Research

Brief History:

The search for the universal language that would uncover the language spoken by biblical Adam, lost at the Tower of Babel, gathered steam in the 17th century by well-known philosophers and mathematicians such as Francis Bacon, René Descartes, Marin Mersenne, Isaac Newton, Gottfried Leibniz, and John Wilkins. It was thought that a system could be discovered that mapped a language-independent symbol to every unique concept. In fact, Bacon believed that Chinese characters were "neither letters nor words" but "things or notions" [Lacr18, p. 163]. Fast forward to the 20th century and we find that many approaches to machine translation (MT) boil down to a similar search for a language-independent universal grammar.

In the 20th century, interest in MT naturally increased as a result of the success of cryptanalysis during World War II and progress in information theory [Shan48]. The beginning of contemporary MT is typically understood to be a result of a memorandum by Warren Weaver [Lacr18, p. 165]. Weaver believed that it was possible to build MT systems that, while not perfect translators, could be within "only X percent error" rate using a common basis between languages. This call for research resulted in 10 years of in-depth investigation and the first MT conference held at Massachusetts Institute of Technology (MIT) in 1952. By 1966, however, the large grants for research from the US government had dried up as a result of slow progress in the field and the lack of evaluation metrics for both MT and human translation.

While some work in MT continued in the 1970s and 1980s, most was commercial with systems focused on weather forecasts, Russian–English translations, and Japanese–English translations. It was not until September 11, 2001 that MT of foreign languages really took off again, with special interest suddenly in Arabic [Koeh10].

Types of Machine Translation:
The two main approaches to MT are rule-based and corpus-based approaches. The first methods were rule-based and were focused on using programmed rules and heuristics for translating a specific language pair. The second approach centers on the idea that rules can be determined automatically from corpora using stochastic processes [Lacr18, p. 171].

Rule-Based Approaches: The three types of rule-based approaches include *direct transfer*, *analysis-transfer-generation*, and *interlingua*. Direct transfer was the earliest approach used and was basically a word-for-word translation technique that used encoded dictionaries. A system designed on this method was used by the US Air Force for Russian−English translation. The analysis-transfer-generation rule-based system breaks MT into three phases. A raw source analysis creates an intermediate representation, for example, a syntax tree, which is fed into a second phase wherein the intermediate representation for the source language is transformed into a similar representation for the target language. This is then used as input to the third phase, wherein the target language translation is produced [Lacr18, p. 173]. Many commercial systems such as Logos, METAL, and Systran used this approach between 1968 and 1990 [Koeh10, p. 16]. The interlingua approach improves upon analysis-transfer-generation by basically eliminating the transfer phase (phase 2). Instead, the intermediate representation is language-independent and can be used to directly generate the target language. Unfortunately, this approach was less promising than the direct transfer approach. It is probably most well known as the approach used in the Universal Networking Language (UNL) Programme, which started in 1996 [UNL20].

Corpus-Based Approaches: Several corpus-based approaches include *word-based statistical MT*, *phrase-based statistical MT*, *syntactic statistical MT*, *and machine learning*. Developed at IBM, and named "Candide," the word-based statistical MT relies on a probabilistic translation dictionary and information theory to learn translation rules from parallel corpora of translated data [Lacr18, p. 171]. Word-based models were then extended into phrase-based statistical models which considered multiple adjacent words and eventually became the dominant approach to statistical MT in the mid-2000s. Google Translate used phrase-based statistical MT for most languages between 2006 [Lacr18, p. 178] and 2016. In the latter year, neural-network-based approaches were shown to be as good, if not better, than the phrase-based method [Monr17]. Syntactic statistical MT combines the best of rule-based and statistical MT systems which have been shown to have state-of-the-art performance for specific language pairs such as English–German. The difference for the rule-based method is that rules are learned automatically from parallel data [Lac18, p178]. Most recently, neural machine translation (NMT), which use neural networks (NNs), have been used for MT and, as stated above, have been shown to outperform the dominant phrase-based statistical approaches. Advances in hardware, the exponential growth in data, and algorithmic improvements have led to the feasibility of NNs for MT. In general, two NNs are trained: an *encoder* for processing the input language and a *decoder* which takes the output in the form of a vector representation of the input and constructs the text in the target language. An attention component is normally used to specifically focus on parts of a source sentence for translation. Adding an attention element to an NMT has been shown to significantly perform better than systems that do not use them [Luon15] and, as a result, there have been additional experiments done to improve NMTs by adding multiple attention mechanisms [Liu19].

Evaluation:

One of the major challenges for MT is how to accurately measure translations. In fact, one of the main reasons that funding for MT research dropped so drastically in 1966 was the lack of good metrics for both human translation and MT. This is important because proper evaluation gives us an indication of the quality of our translations.

The gold standard for evaluation of translations is to use humans fluent in both the original language and the translated language. But even this is not fool proof since two humans fluent in the same languages may not translate the same source into identical translations. We can create metrics around inter-annotator agreement (how well two or more translators can make the same decision for a certain category) [Koeh10], but this is expensive both in terms of time and personnel.

For many reasons (time, expense, adaptability, etc.), automatic methods for evaluation are preferred. The most common automatic metric is called BLEU, or Bilingual Evaluation Understudy [Papi02]. The BLEU metric has been shown to highly correlate with human judgment for translation and is the usual metric most automatic evaluation methods use to demonstrate the effectiveness of their own systems.

4.2.2 Summarization as a Means to Functionally Grade the Accuracy of Translation

Functional assessment of translation is important since it is difficult, time-consuming, and often quite expensive to find human ground truth capabilities (people fully fluent in the two or more languages). Thus, summarization and its counterpart, query set generation, can be used to quantitatively grade the accuracy of multiple translation engines. As usual in this book, we are primarily concerned with providing a *functionally equivalent* representation of content that can be used to perform the same task as the original content. Summarization is the usual means of reducing a larger set of content to a functionally equivalent set of content, with the risk that the same summarization will not be optimum for later tasks, unanticipated by the original summarization approach. However, translation is one of the few text analytic tasks that are unlikely to suffer from this "de-optimization" effect. This is because the summarization almost certainly must occur in the *source language* in order to reduce the variance associated with a task (summarization) that is highly susceptible to noisy input. Summarizing after translation suffers the problem of matching idiomatic expressions between the source and destination languages. The disparity in how idiomatic expression affects word frequencies and rareness of words is one reason why the summarization should be performed in the source, and not the destination, language. For example, the expression "let's get on the same page" has its equivalent in French of "accordons nos violons," which involves relatively rarer words "accordons" and "violons" than the six very common words in the English expression. Thus, this prosaic English expression is unlikely to be chosen for an extractive summarization in English, but "accordons nos violons" has a much higher chance of being chosen in French due to the relative rarity of its terms.

There is an optimal percentage of the document (often between 10% and 20% of medium length documents) used in a summarization which, when analyzed, leads to the most equivalent behavior to the original text in a variety of text analytic tasks. Summarization that is used in its translated form is, of course, determined in the original language to prevent the types of variances described in the preceding paragraph. However, summarization itself can be used to **functionally grade the accuracy of translation**. It is accomplished in the following fashion. The set of documents of interest, along with a set of queries (searches) to be performed on the documents, are translated. For both the original set and the translated set, the relationship between query terms, or [query set}, and the document set, or {document set}, are noted. The relationship between queries and documents is used as a measurement of the corpus behavior. That is, the best translator is deemed to be the one that provides a translated {query set, document set} that behaves the most similarly to the {query set, document set} of the original language text. In this functional test, summarization is an auxiliary function, not optimized for its own usual utility (that of providing the best comprehension of a document in the minimum number of words) but, instead, as a further means for measuring the robustness of translation. Summarized forms of the document provide a middle layer between the large documents and the small queries, and we denote this as the {summarization set}. Now, we can compare the links between the {query set, summarization set} in addition to the {query set, document set}. Perhaps, the most straightforward manner in which to perform this comparison is to use the Percent Matching Queries, as defined by Equation (4.1).

$$\text{Percent Matching Queries} = \frac{(\text{Number of Matching Queries})}{(\text{Number of Queries})}. \quad (4.1)$$

Since the number of queries is identical for both languages, and the number of matching queries is readily computed by the number of queries which return the same summarization (i.e., the summarization of the same document), Equation (4.1) is essentially the percentage of queries (PMQ) which are exact matches in the second language (assuming the matches in the first language are exact, and thus for our purposes "exactly correct"). Equation (4.1) is applied to either the {query set, summarization set}, for which it is designated PMQ(q,s) or to the {query set, document set}, for which it is designated PMQ(q,d). These two results can be used together to find PMQ(total) according to Equation (4.2). Here, the multiple λ is a value between 0 and 1, and so PMQ(total), like PMQ(q,s) and PMQ(q,d), is also a value between 0 (a terrible translation, we assume) and 1 (a perfect translation, presumably).

$$\mathrm{PMQ}\,(\mathrm{total}) = (1 - \lambda)\,\mathrm{PMQ}\,(q, s) + \lambda\mathrm{PMQ}\,(q, d)\,. \qquad (4.2)$$

This is not the only possible means of scoring the query match for either {query set, summarization set} or {query set, document set}. For example, instead of only noting exact matches, we may wish to note when the query returns the correct document as the 2nd, 3rd, 4th, or other relatively highly ranked match. We then apply a ranking factor R(match) which might be 1.0 for an exact match, 0.8 for 2nd best match, 0.6 for the 3rd best match, etc. In general, this leads to Equation (4.3).

$$\mathrm{Ranked\ Matching\ Score} = \frac{\sum_{i=1}^{\mathrm{Number\ of\ Queries}} \mathrm{Rank\ of\ Match}(i)}{(\mathrm{Number\ of\ Queries})}. \qquad (4.3)$$

If we denote the ranked matching score in Equation (4.3) as RMS, then we can proceed with RMS(q,s) and RMS(q,d) in the same manner as for PMQ in Equations (4.1) and (4.2). The general form of RMS is therefore given in Equation (4.4).

$$\mathrm{RMS}\,(\mathrm{total}) = (1 - \lambda)\,\mathrm{RMS}\,(q, s) + \lambda\mathrm{RMS}\,(q, d)\,. \qquad (4.4)$$

The approach described here was originally proposed, in abbreviated form, in an earlier work [Sims14]. This brief section has provided the quantitative approaches necessary to enable such a system. We next turn to the machine learning aspects of language translation.

4.3 Machine Learning Aspects

4.3.1 Summarization and Translation

The previous section outlined a functional machine learning approach to language translation, wherein the steps of the algorithm to rank order multiple translators is as follows:

(1) translate document set D(L1) in language L1 into document set D(L2) in language L2;
(2) translate query set Q(L1) in language L1 into document set Q(L2) in language L2;

(3) translate summarization set S(L1) in language L1 into summarization set S(L2) in language L2;

(4) determine the best translator as the one maximizing Equation (4.2) or Equation (4.4).

Functional means of assessing translation have a lot of benefits. In the next section, we will illustrate how expertise in curriculum development can be used to select the optimum translator. It is important to note in these discussions that the "optimal translator" may be selected from a population of different translators, but it can just as readily (and appropriately) be selected from among a population of the same translator being used with different settings. In other words, we can optimize a given translator as easily as we can select from among a plurality of distinct translators.

4.3.2 Document Reading Order

Another functional means of assessing language translation accuracy is by comparing the reading order of a document collection. In previous work [Kout15], we showed that for a set of documents, a specific reading order may be suggested to the readers based on their relevance to the user's query. Thus, reading order is simply the ranked order of documents returned as a result of a query. Such an approach is well-suited to finding a needle in a haystack – that is, finding the specific document one is looking for from a large corpus. However, this approach is not ideal when a user needs to access germane documents in some logical order, as in, for example, when learning or when editing. One means of generating a reading order is to organize the collection of documents of interest in a tree arrangement, with the more general documents at the top and the more specific documents at the bottom of the tree. The user can then select a path from the top to the bottom. This path comprises the **reading sequence** over the documents. This relatively novel way to provide sequential content consumption departs from the typical ranked lists of documents based on their relevance to a specific user query [Kout15] and is also quite different from a static navigational interface.

Different document orders can be compared directly, using a modified Hamming Distance to compute the similarity between two reading orders. Suppose there are 10 salient documents which, for a hand-picked (ground truthed) reading order, should be read in the order as given in Equation (4.5); that is, from D1 to D10:

$$\{D1, D2, D3, D4, D5, D6, D7, D8, D9, D10\}. \qquad (4.5)$$

Then, suppose that a given reading order algorithm suggests that the document be read in the order indicated in Equation (4.6):

$$\{D3, D4, D1, D6, D2, D7, D8, D5, D9, D10\}. \qquad (4.6)$$

The question is then, how similar are these two reading orders? In this section, we will provide three measures of reading order similarity. In two cases, a lower score implies higher similarity, and in the other case, a lower score.

The first two measures, or metrics, are based on the sum of the order errors in the two lists. The first is based on the order errors alone and so is designated the **order error**. For example, if D2 is in the second spot in one reading order list and in the fifth spot in another reading order list, then the order error is 3 for this document. In comparing the two lists in Equations (4.5) and (4.6), then, the order errors are (2, 3, 2, 2, 3, 2, 1, 1, 0, 0) for (D1, D2, D3, D4, D5, D6, D7, D8, D9, D10). For example, "D2" was in the 2nd position in Equation (4.5) but the fifth position in Equation (4.5); so its order error is 3, the absolute value of the difference between {5} and {2}. The sum of these order errors is 16. The **mean order error** is thus 1.6. The second of the two order-based approaches assigns weights more highly to the documents that are higher in the order list. So, the order error is multiplied by a factor inverse to the rank, e.g., (10, 9, 8, 7, 6, 5, 4, 3, 2, 1) for ranks $1-10$, respectively. This **weighted order error** is (20, 27, 16, 14, 18, 10, 4, 3, 0, 0) when multiplying (2, 3, 2, 2, 3, 2, 1, 1, 0, 0) × (10, 9, 8, 7, 6, 5, 4, 3, 2, 1) for comparing the results in Equations (4.5) and (4.6). The weighted order error is the sum of these 10 values, or 112. The **mean weighted order error** is 11.2. These values are included in Table 4.2.

Another metric for similarity between the ground truthed reading order and the experimentally determined reading order is the (linear) correlation coefficient. This is illustrated for comparing the reading order "equations" Equation (4.5) and Equation (4.6) as shown in Figure 4.1. The correlation coefficient, or R^2 value, is 0.6112 for the linear correlation, implying that over 61% of the variability in the experimental reading order is accounted for by its correlation with ground truthed reading order. This value is also tabulated in Table 4.2.

A separate algorithm is used to generate a reading order for the 10 documents in Equation (4.5). This new "experimental" reading order is given in Equation (4.7):

$$\{D1, D4, D3, D5, D2, D8, D6, D7, D10, D9\}. \qquad (4.7)$$

Table 4.2 Similarity values when comparing Equations (4.6) and (4.7) to Equation (4.5). Please see text for details. In each case, the reading error results for Equation (4.7) are closer to ground truth than the reading error results for Equation (4.6).

Similarity values	Equations (4.5) and (4.6)	Equations (4.5) and (4.7)
Order error	16	12
Mean order error	1.6	1.2
Weighted order error	112	63
Mean weighted order error	11.2	6.3
Correlation coefficient	0.6112	0.7511

The **order error** set for Equation (4.7) compared to Equation (4.5) is (0, 2, 0, 1, 3, 2, 1, 1, 1, 1) for (D1, D2, D3, D4, D5, D6, D7, D8, D9, D10). The sum of these order errors is 12. The **mean order error** is thus 1.2. The **weighted order error**, computed as above for comparing Equation (4.6) to the ground truthed Equation (4.5), is (0, 18, 0, 7, 18, 10, 4, 3, 2, 1), obtained by multiplying (0, 2, 0, 1, 3, 2, 1, 1, 1, 1) × (10, 9, 8, 7, 6, 5, 4, 3, 2, 1). The weighted order error is the sum of these 10 values, or 63. The **mean weighted order error** is 6.3. The plot of the reading order "equations" Equation (4.5) and Equation (4.7) is shown in Figure 4.2. The correlation coefficient, or R^2 value, is 0.7511 for the linear correlation, implying that over 75% of the variability in the experimental reading order is accounted for by its correlation with ground truthed reading order. These values are included in Table 4.2. For each metric, it is clear that the reading order for

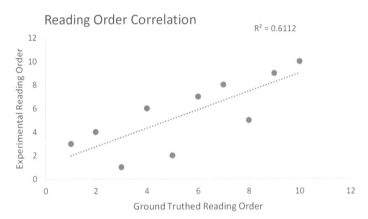

Figure 4.1 Reading order correlation between the experimentally determined reading order (Equation (4.6)) and the ground truthed (human evaluated) reading order (Equation (4.5)). The correlation coefficient, R^2, is 0.6112, meaning 61% of the variance in the experimental reading order is due to its correlation with the ground truthed reading order.

Equation (4.7) is closer to the human labeled reading order of Equation (4.5) than the reading order of Equation (4.6).

This short example illustrates a simple approach to relatively comparing machine learning output for reading orders. More importantly for the topic of this chapter, if the reading orders obtained for Equations (4.6) and (4.7) are computed from two different translations of the document set D1−D10, which had the suggested reading order in Equation (4.5) when analyzed in the source language, then we can functionally declare the second translation (resulting in Equation (4.7)) to be superior to the first translation. Some may be uncomfortable with this approach to measuring the relative effectiveness of different translation engines. However, we believe that this functional measurement is actually a significant improvement over traditional ground-truthing-based approaches for at least the following reasons:

(1) It is not arbitrary: The use of order error, weighted order error, and the correlation coefficient means that the assessment is performed automatically once the reading orders have been determined. The comparison is entirely mathematical, and as such the results are not biased toward the particular preferences of one or more of the humans otherwise scoring the documents.

(2) It is readily scalable and, effectively, "infinitely scalable." Since the scores calculated for the reading order matching can be added for any number of reading order experiments, the translators can be compared for as large a data set as desired. In fact, multiple runs of the translators

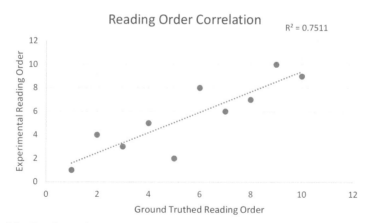

Figure 4.2 Reading order correlation between the experimentally determined reading order (Equation (4.7)) and the ground truthed (human evaluated) reading order (Equation (4.50)). The correlation coefficient, R^2, is 0.7511, meaning 75% of the variance in the experimental reading order is due to its correlation with the ground truthed reading order.

may create a "sample set" of reading order metrics, which can then use traditional parametric statistics (e.g., analysis of variance) to show statistically significant differences between translators.

(3) Only one domain expertise-driven ground truth entry, the preferred reading order in the source language, is needed for the system. This is a huge benefit for the overall system. It means, for example, that proficiency in only one language is required. It also means that, in many cases, *no ground truthing is required* because the germane reading order already exists in the form of a curriculum for a given subject area.

The third reason is particularly important and is directly related to the primary justification for this book. When it comes to ground truth, less is better. Hand-crafted training of systems is not only hard work, but it can age quickly and, thus, have only evanescent relevance. More insidiously, the goals of the ground truthing agents might be divorced from the current or future goals of the objects ground truthed. Ground truthing personnel might be interested in finding more information of a certain type than another and so be "biased" in what they are looking for. This may be okay for the original intent of the ground truthing, but if and when the same ground truthed information is used for an alternative purpose, the difference in the rate of false positives for the original information and the rate of false negatives for the later-desired information will affect the overall utility of the ground truthing.

4.3.3 Other Machine Learning Considerations

Accurate translation, of course, requires a lot more machine learning than we are going to cover here. Languages like English contain over 100,000 words in the dictionary, some of which have dozens of entries. And, these are just the single words. Most of the richness of a language derives from compound nouns, compound verbs, idiomatic expressions, neologisms, plays on words, slang, jargon, and other meta-definitional words and phrases. In English, many speakers likely have at their avail millions of different terms, compound terms, and idiomatic expressions. As one example, condition the future progressive tense, as in I "will be doing" something. A sentence containing two examples of the future progressive is "tomorrow I will be swimming since tomorrow I will be visiting the beach." This illustrates the need for very large training sets for a language-based system such as a translation engine. Supposing English or another highly complicated modern language (Mandarin, French, Hindi, Spanish, Russian, etc.) has one million different terms of differential meaning/usage, our ground truthing may need to consist of 100 million words just to ensure that 99% of these expression occur in the ground truth. This is a rather unwieldy data set for everyday purposes.

In short, then, the massive training demands associated with applying machine learning to text analytics systems argue for a different means of generating the training sets. In this section in specific, and this book in general, we argue for the identification and adoption of functional means to provide an effective training set. Not only is this "easier," but it may also be more robust.

4.4 Design/System Considerations

Language is often – sometimes sententiously, but almost certainly still accurately – portrayed as the main driver of human advancement and differentiation from other intelligent creatures. What brings sapience to the cerebrum of *Homo* as opposed to the advanced brains of many other primates and cetaceans that might be candidates for intellectual primacy on the planet? Right, it has got to be language. One might comment that the architecture of the human brain seems well-suited to language, but that could well be *a posteriori* thinking or at minimum overtraining the data model. Not to mention the fact that the architecture of the brains of other intelligent species, from dogs to dolphins and from orcas to otters, are largely homologous to the human brain architecture. Looking more closely, there is nothing in the nature of understanding our universe – and internalizing it into a consciousness – that requires auditory information to pass from the temporal lobe to the parietal lobe; visual information from the occipital lobe to the parietal; and somatosensory information from the frontal lobe and anterior parietal lobe to join the incoming audiovisual information where it does. Nor is there any Platonic universal form requiring vestibular senses to be associated with auditory senses, nor for olfactory and gustatory senses to enter the brain so much closer to the spinal cord. It is that way in all intelligent animals *on this planet* because intelligence was fought long and hard for, in a personification of the process of natural selection. Once intelligence was garnered in an evolutionarily successful species, the road map for intelligence was passed preserved to the offspring or the offspring had no hope of surviving.

From this perspective, then, the neural architecture necessary to engender advanced intelligence on Earth is a local optimum. Those familiar with evolutionary algorithms know that optimization often settles on a non-global optimum simply because the neighborhood around this not-truly-optimum is much larger than the neighborhood around the global optimum. This is akin to walking into a movie video store in those halcyon days of the 1990s and settling on "Titanic" or "Jurassic Park" instead of the arguably far more compelling "Fargo" or "Malcolm X" just because the neighborhood around them is so much bigger (100 copies of "Titanic," 80 copies of "Jurassic Park," but only one each of "Fargo" and "Malcolm X")! Maybe that analogy was pushed too far, but the point is well taken: the optimal solution among all possible solutions, Dr. Pangloss' expected outcome of any optimization routine, may be a palace hidden in a rather dicey neighborhood.

Further countering the, we feel, false notion that the human brain is perfectly designed for language, let us consider its architecture further. It is safe to say that if the human brain were designed to process language in its spoken, heard, read, and even haptic (e.g., Braille) form, the brain might be designed to place these disparate inputs and outputs closer together at the thalamus level (that is, in the diencephalon, the very heart of the brain), and not in the parietal lobe. This architecture, had it been the product of an intentional design rather than the survivor of the vicissitudes of evolution, would allow a lot more crossover of content earlier in the language processing timeline. The fact that dyslexia, apraxia, agnosia, agraphia, alexia, aprosodia, and dysphemia – to name just a few of the dozens of language disorders – are all relatively common in humans is a fair, if anfractuous, argument for the architecture of the human brain not being the global optimum design. Also − it is important to note that language is only one of the many design constraints driving the overall architecture of the brain. Further elaboration on this multi-constraint optimization problem is the appropriate subject of a different book, and we will not push further here except to say that the architecture of the brains of intelligent species on this planet are the product of a multiplicity of optimization functions being applied simultaneously.

Why is it important to understand the limitations of human language processing? For the same reason that it is important to understand the limitations of human-designed, machine-learning-based, language processing. As with most, if not all, machine-learning-based systems, tradeoffs are essentially to make the system meet a wide variety of desired specifications. Among the specifications that must be considered is the following set of five examples:

(1) The system must be able to understand N different languages compromising the set $\{L1, L2, L3, \ldots, LN\}$.
(2) The system must be able to perform its linguistic tasks through some combination of local processing (e.g., on a mobile phone, pad, or laptop) and cloud-based processing (when there is sufficient bandwidth).
(3) The system must be able to understand relevant dialects (and accents) within the set of languages it supports.
(4) The translation (or other linguistic metric) accuracy must exceed $A\%$.
(5) The process must complete in less than T seconds.

To this list, you can surely add a few more examples of your own. The key aspect of this is that once you have two or more specifications, they generally constrain one or more of the other specifications. For example, higher accuracy generally requires more overall processing. To meet *A%* accuracy, we may not be able to complete our text analytics in under *T* seconds. Constraints can become complaints, and to avoid those, we are often incented to architect systems with a locally optimum design but do not contradict the constraints.

In many applications, then, a specific set of system settings are locked in place early in the design. These settings may not be optimal, but they may be reliable, familiar, or even required based on some standard, business practice, or even governmental or regulatory policy. The problems with such a non-global optimum are well-known: they are not robust to new data, they do not adapt to the changing needs of their users, and in some circumstances, they may even be dangerous. However, short of having massive volumes of training data representative of every subdomain of the input's domain, some form of compromise is almost always made during optimization. After all, *exhaustive search is not optimization*, it is simply the output of a full enumeration: it is like being asked to guess which US state has the highest population and requesting 50 guesses. That is a nice luxury to have, but it is fundamentally flawed, even in a circumstance when some technology – say quantum computing – brings near-infinite computing power to your problem. The flaw is that language requires making a choice. You do not get to try five different answers to your partner's question, "What's on your mind?" You must instantly choose "I'm worried about the future of the planet" or "I'm thinking of going for an ice cream." Once the first choice is made, the conversation will be steered (usually, anyway) along a different branch of the decision tree of conversation. Your partner's response operates under the same set of constraints: a plethora of choices, only one of which will be thrown into the unknown future of the conversation. Once a conversation has been completed, the sequence of utterances might make sense, but this does not imply it could have been predicted beforehand.

The argument against exhaustive search for linguistic applications is also an argument for functional means of determining the output. Individual expressions in a conversation, after all, are very difficult to judge in isolation. If you were to say, "That's great!" you may actually mean it, or you are – perhaps more likely – being sarcastic. The most important means of analyzing conversational language (and other language, in fact) is in the context around the conversation. This context is not just important, it is essential for any form of comprehension. Internalizing this observation, then, we wish to reconsider the manner in which training is provided for text analytic systems. Instead of focusing on optimization, the focus should be on fluency. Language should flow, not at some pre-determined or pre-imagined optimum velocity, but without damming, without diversion into some form of linguistic Faiyum, and always supporting its craft. When you have mastered a language, you do not hold a huge vocabulary behind a philological reservoir, you do not provide grammatical perfect parsing – no, you are considered *fluent*.

From this viewpoint, training data, at least for conversational linguistics, should be tied to making the communication fluent. Unfortunately, in many text analytic systems, more traditional measures of accuracy, performance, statistically significant differences between groups, and other concerns still prevail. We break with them here. Two simple, but novel to non-linguistic machine learning systems, approaches are instead discussed. These are (1) the use of dialect to test the robustness of a system, and (2) the use of a "telephone" game to provide the means for assessing the robustness of language-as-message.

The first of these is often overlooked. In spite of research extending back several decades assessing the error rate for different broad accents [Yan02], variations in accents are not routinely employed in the assessment of text analytics. In our view of system design, the training set should be as carefully balanced for the accents of its users as it is for age, gender, race, and economical status distributions. Accent accuracy may provide a means of predicting translation accuracy. We consider this through an example that performs the following steps:

(1) A speech-to-text engine that can be used to convert audio streams of sentences spoken in a specific language to a string of text that can be compared to the ground truth.
(2) Two different translation engines (speech-to-speech), each of which can translate the audio stream in one language to an audio stream in another language.
(3) The audio stream in the second language in (2) above is then converted to text that can be compared to the ground truth.

These output of applying (1), (2), and (3) with two different translation engines are shown in Figures 4.3 and 4.4. In each of these figures, the accuracy for correctly understanding a set of 100 sentences (randomly assigned from a set of 200) in each of 10 English accents (x-axis) is plotted on the x-axis (this uses the speech-to-text engine). A sentence is considered correctly understood when all of the words in the sentences are correctly output by the speech-to-text software. If one or more words are misidentified, the sentence is considered inaccurate. For the 10 accents, between 34 and 81 sentences are correctly understood for Translation Engine 1 (Figure 4.3) and between 33 and 87 sentences are correctly understood for Translation Engine

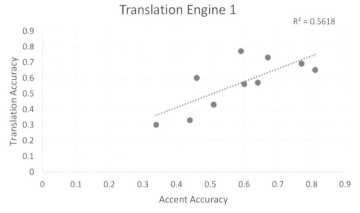

Figure 4.3 Translation accuracy versus accent accuracy for Translation Engine 1. The mean accent accuracy is 0.583, with a mean translation accuracy of 0.563. The correlation coefficient $R^2 = 0.5618$, meaning there is a strong positive correlation between the two accuracies.

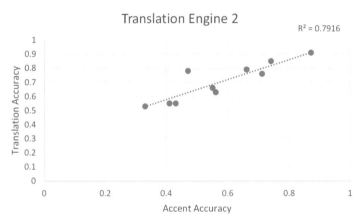

Figure 4.4 Translation accuracy versus accent accuracy for Translation Engine 2. The mean accent accuracy is 0.573, with a mean translation accuracy of 0.701. The correlation coefficient $R^2 = 0.7916$, meaning there is a very strong positive correlation between the two accuracies.

2 (Figure 4.4). The means of the two engines, as should be expected, are very similar: 58.3 sentences for the 10 accents in Figure 4.3, and 57.3 sentences for the 10 accents in Figure 4.4. The y-axis for each of these figures represent the translation accuracies of the words in English into another language (here French), for which Translation Engine 1 had a range of 0.30–0.77 (mean = 0.563), and Translation Engine 2 had a range of 0.53–0.91 (mean = 0.701). The first Translation Engine (Figure 4.3) has an $R^2 = 0.5618$, meaning there

is reasonably strong positive correlation between the accent understanding accuracy and translation accuracy. The second Translation Engine (Figure 4.4) has an $R^2 = 0.7916$, meaning there is a much stronger positive correlation between the accent understanding accuracy and translation accuracy.

From Figures 4.3 and 4.4, we do not see the accent accuracy as a predictor for translation accuracy. Since the same speech-to-text software was used for both experiments, we expected the accent accuracies to be identical if the two data sets are similar. That was achieved, as accent accuracy was a mean of 0.583 for Figure 4.3 and 0.573 for Figure 4.4. However, the translation accuracies behave differently, with that of Figure 4.3 (mean translation accuracy = 0.563) being considerably less than that of Figure 4.4 (mean translation accuracy = 0.701). Higher translation accuracy corresponds to a higher correlation coefficient. Thus, a higher correlation between accent accuracy and translation accuracy is, in this example, an indicator of better translator engine accuracy. One might be tempted to ask, why not just perform the speech-to-speech text process, that is, steps (2) and (3)? Certainly, this shows which translation engine has the higher accuracy. However, we may not be able to compare the same sets of input for the two engines (and this was definitely the case here, although the random 50% of the training data assigned to the two experiments was clearly very similar). Being able to use any speech data set and simply compute the correlation between accent and translation accuracy provides a good measure of system robustness. A more robust language system is going to perform accent understanding and translation with similar accuracies and, thus, has a higher correlation coefficient. In this experiment, we are right in thinking Translation Engine 2 is more accurate than Translation Engine 1 on the basis of the R^2 values. The fact that the translation accuracy was also computed shows this is correct, but even had the two translation accuracies been similar, the higher R^2 value for Translation Engine 2 would have led us to favor it over Translation Engine 1.

The second of the two simple approaches to assessing translation accuracy is the use of the "telephone" game process. The "telephone" game is ideal for an icebreaker, party, or other casual get together. In this game, the players form a circle. The first person is given a message, which they whisper into the ear of the person next to them. That person passes the message to the person on their other side, and this continues until the message comes back to the last person in line (e.g., to the left of the message creator if the messages are passed to the person on the right). The original and final messages are compared to one another, and the word distance between the two sentences can be computed. We recommend the use of a modified form of the Damerau−Levenshtein distance [Dame64][Leve66], well-known for individuals words. The Damerau−Levenshtein distance is the minimum number of insertions, deletions, and substitutions of a single item (these three add to the Levenshtein distance) along with transposition of adjacent items (added to the Levenshtein distance, this becomes the Damerau−Levenshtein distance) required to make the two lists identical. For words, these items are letters; for sentences, they may be words. So, to compare the word "steaks" to the word "stakes," we have a Levenshtein distance of 2, e.g., delete the "e" to get "staks" and then reinsert the "e" after the "k" to get "stakes," and Damerau distance of 2, e.g., transpose the "e" and "a" to get "staeks" and then transpose the "e" and "k" to get "stakes." No combination of the four operators gives a distance less than 2, so that is our Damerau−Levenshtein distance. For a sentence such as "Bob will pay the bill and leave" to change to "Bob and Will, pay the bill" we delete the word "leave," delete the word "and," and insert the word "and" and get "Bob and Will pay the bill." Ignoring the punctuation, this is a Levenshtein distance of 3 and shows how a quantitative score can be given to differences in the original and final messages in a game such as "telephone." For the purposes of evaluating a translator, we may wish to chain the translations back and forth between two languages, with each pair of translations – e.g., from Spanish to Mandarin and Mandarin to Spanish – constituting a single "round trip." The degradation of the message with round trip is a form of "telephone rating" for the translator. Suppose, in the above example, the Damerau−Levenshtein distance for the 1000 original words in each experiment are {24, 31, 40, 43, 48} for Translation Engine 1 and {18, 24, 27, 33, 34} for Translation Engine 2. These data again support the interpretation that Translation Engine 2 is more robust since it has a lower Damerau−Levenshtein distance after 1, 2, 3, 4, or 5 round trips. Other useful functional approaches to assessing translations associated with telephone-like approaches will be discussed below.

4.5 Applications/Examples

The applications of language translation are as diverse as language itself. In being able to interconnect two different language systems, translation is in many ways the jewel in the crown of text analytics. Incorrect translations can be security and/or privacy risks, not to mention embarrassing to the original speaker. From a functional text analytics standpoint, translation accuracy on a ground truthed training set can be used as a functional measurement for the various text analytics required to produce the translation, including the various elements in natural language processing (NLP) such as a part of speech (POS) tagging, compound word identification, and the like. Any improvement in computed translation accuracy when replacing one

Table 4.3   Table of translation accuracies between the five "EFIGS" (English, French, Italian, German, and Spanish) languages for a particular translation system.

FROM↓/TO→	English	French	Italian	German	Spanish
English	1.000	0.951	0.953	0.968	0.967
French	0.943	1.000	0.902	0.878	0.911
Italian	0.944	0.907	1.000	0.873	0.916
German	0.961	0.867	0.884	1.000	0.833
Spanish	0.964	0.902	0.923	0.856	1.000

component with another in the overall process can be viewed as a measure of the relative functional value of the component.

Perhaps not surprisingly, translation engines are of both **non-uniform** and **asymmetric** quality for translating. This is illustrated in the data collected in Table 4.3 for a translation engine that can translate between any pair of "EFIGS" (English, French, Italian, German, and Spanish) languages. In Table 4.3, it is clear that the "central" language for the translation engine is English, with a mean translation accuracy of 0.960 into the other four languages, much higher than the (0.909, 0.910, 0.886, 0.911) values for (French, Italian, German, and Spanish), respectively. Also, it is much more accurate to translate into English (mean = 0.953) than into French (mean = 0.907), Italian (mean = 0.916), German (mean = 0.894), or Spanish (mean = 0.907). This translation engine is therefore most likely designed to help a person whose primary language is English interact with content in other major European languages.

The dependency of the translation engine on English is made even clearer by Table 4.4, in which it is shown that in 2/3 of the cases, translating between two non-English languages is more accurate when using translation of the source language into English, and then from English into the destination language, than simply performing the source-to-destination translation. The data also show that, in general, Italian is the next most proficient language of the designers of the translation engine since it directly translates into French and Spanish with higher accuracy than when using English as an (extra) intermediate translation.

For another functional application of translation and, in particular, the "telephone" inspired sequential or "chained" translations, we return to a discussion of privacy and security. In a Markov chain with N nodes, previous work [Sims19] has shown empirically that memory of a specific previous state does not last for more than N state transitions. One advantage of the "telephone," or chained translation, approach, then, is to obfuscate the source of the content. This could be desired for several reasons, including protecting a whistleblower, protecting a journalist, or protecting a "secret shopper" gaining evidence of illicit commerce. It can also be used to ensure that the

Table 4.4 Table of translation accuracies (range 0.00–1.00) between the four non-English languages "FIGS" (French, Italian, German, and Spanish) languages when using translation into English first as an intermediary. Values in ***bold italics*** are more accurate than direct translation as in Table 4.3. This is the case for 8 out of the 12 possible translations. The diagonals (0.897, 0.900, 0.930, 0.932) are the single round-trip accuracies for the combination of English and (French, Italian, German, and Spanish), respectively.

FROM↓/TO→	French	Italian	German	Spanish
French	0.897	0.899	*0.913*	*0.912*
Italian	0.898	0.900	*0.914*	0.913
German	*0.914*	*0.916*	0.930	*0.929*
Spanish	*0.917*	0.919	*0.933*	0.932

Table 4.5 Confusion matrix of results for classifying 1000 documents each for four source languages (French, Italian, Portuguese, and Spanish). The documents are 100% accurate in classification based solely on a dictionary matching algorithm. Most importantly, this table illustrates how documents from different languages (different rows) are assigned to different languages (different columns). Perfect classification results in non-zero elements along the diagonal only, as in this matrix.

Language confusion matrix		Destination language (classification result)			
		French	Italian	Portuguese	Spanish
	French	1000	0	0	0
Source language	Italian	0	1000	0	0
	Portuguese	0	0	1000	0
	Spanish	0	0	0	1000

translations are consistent across all of the languages. Another interesting aspect of this form of "source anonymization" is that we can provide a "homogenization" of semantics, or meaning, among the languages after moving through a sufficient number of "round trips" of translation. This will be illustrated in an example using a confusion matrix approach, which is a matrix that provides the source language on one axis and the classification of the source on the other axis. The first confusion matrix shown is that for documents that are not translated. This is provided by Table 4.5, and as this table shows, 100% of the documents from the four languages (here, French, Italian, Portuguese, and Spanish are represented by 1000 documents each) are correctly classified as coming from those languages since they have not yet been translated. A classification approach as simple as dictionary matching (used here) provides 100% accuracy since the four languages have significantly different vocabularies.

Table 4.6 provides a normalized version of the confusion matrix in Table 4.5. This is the normal form of the confusion matrix which will therefore be used in the rest of this section.

Table 4.6 Normalized version of the confusion matrix in Table 4.5. Perfect classification results in non-zero elements along the diagonal only, as in this matrix.

Language confusion matrix		Destination language (classification result)			
		French	Italian	Portuguese	Spanish
	French	1.00	0.00	0.00	0.00
Source language	**Italian**	0.00	1.00	0.00	0.00
	Portuguese	0.00	0.00	1.00	0.00
	Spanish	0.00	0.00	0.00	1.00

In this section, we are concerned with the confusion matrices not as much for classification accuracy as we are for source language (or "origin") identification, or recognition. In Table 4.7, we show the results for attempting to identify the original language of each of the 4000 documents after they have been translated into English. As we can see from the results, the source language is identified in 86.8% of the instances (summing the diagonals and dividing by the number of languages gives the value 0.868 for the normalized confusion matrix of Table 4.7). Since this value is substantially above the 0.250 value expected by random guessing, it is clear that the translation engine has some language–language dependencies and creates a different version of English for each of the four source languages.

For the confusion matrix shown in Table 4.7, in order to obfuscate the source language of the document, we should arrive at a confusion matrix with the following characteristics:

(1) Each of the rows should sum to 1.000 since the confusion matrix is normalized.
(2) All elements of the confusion matrix should converge on $1/N$, where N is the number of source languages. Here, $N = 4$ and so we expect all $[i,j]$ $\rightarrow 0.25$ as the source language is successfully obfuscated.

Table 4.7 Accuracy of identification of the source languages for documents translated into English from one of the four FIPS (French, Italian, Portuguese, and Spanish) languages. Since the diagonal elements (mean = 0.868) are well above the value of 0.250 for random guessing, it is clear that the source language is not de-identified by the translation engine since the correct source language is identified in 86.8% of the instances (3472 out of 4000 documents).

Language confusion matrix		Destination language (classification result)			
		French	Italian	Portuguese	Spanish
	French	0.852	0.063	0.035	0.050
Source language	**Italian**	0.029	0.887	0.041	0.043
	Portuguese	0.033	0.077	0.859	0.031
	Spanish	0.039	0.044	0.043	0.874

(3) The trend in (2) can be directly measured by the standard deviation of the values in the confusion matrix. When full de-identification of the source language has occurred, the standard deviation will be small and will be the same for the on-diagonal and off-diagonal elements.

We next translate the documents into one of the non-English documents and retranslate them into English. So, for the Italian-source documents, 250 each of the English-language versions created for Table 4.7 are translated from English into French; another 250 into Italian (back to Italian for these); another 250 into Portuguese; and the final 250 into Spanish. These documents are then translated back into English and we compute a new confusion matrix, as shown in Table 4.8.

The results of Table 4.8 show that the additional round trip of translation has helped de-identify the source language. Random guessing would result in the source of 1000 documents being correctly identified. In Table 4.7, 3472 documents were correctly identified; in Table 4.8, only 1394 documents were correctly identified. This means that the extra round trip has de-identified $|3472 - 1394|/|3472 - 1000|$, or 0.841, of the remaining above-expectation identification in the document corpus. One final round trip is added in Table 4.9, where a document is round trip translated from English into one or the other of the remaining languages and back into English. So, if a French-source document had been translated into Italian in Table 4.8, it is now either translated into Spanish ($n = 125$) or Portuguese ($n = 125$) and back into English.

The results of Table 4.9 are compelling. After the second additional round trip, the number of correctly identified source languages is only 34 (0.85%) above random guessing. This means overall that the mean percent identified for each source language is 25.85%, and the mean percentage obfuscated

Table 4.8 Accuracy of identification of the source languages for documents translated into English from one of the four FIPS (French, Italian, Portuguese, and Spanish) languages, then round-trip translated in equal sized groups back into one of the FIPS languages and then again into English. This constitutes an extra "round trip" of translation over Table 4.7. Here, the diagonal elements (mean = 0.349, 1394 out of 4000 documents are in the diagonals) are much closer to the value of 0.250 for random guessing in comparison to the value of 0.868 of Table 4.7 for the single translation case. Here, the source language has been de-identified much more successfully.

Language confusion matrix		Destination language (classification result)			
		French	**Italian**	**Portuguese**	**Spanish**
	French	0.347	0.232	0.197	0.224
Source language	**Italian**	0.197	0.359	0.209	0.235
	Portuguese	0.176	0.299	0.339	0.186
	Spanish	0.185	0.237	0.229	0.349

Table 4.9   Accuracy of identification of the source languages for documents translated into English from one of the four FIPS (French, Italian, Portuguese, and Spanish) languages, then round trip translated in equal sized groups back into one of the FIPS languages and then again into English, and then into one of the two remaining FIPS languages back into English. This constitutes two extra "round trips" of translation over Table 4.7. Here, the diagonal elements (mean = 0.259, 1034 out of 4000 documents are in the diagonals) are much closer to the value of 0.250 for random guessing than either Table 4.7 or Table 4.8. The residual identification of source language is on the order of about 1.13% (the difference between the mean of the diagonals and the mean of the off-diagonals).

Language confusion matrix		Destination language (classification result)			
		French	**Italian**	**Portuguese**	**Spanish**
	French	0.255	0.249	0.248	0.248
Source language	**Italian**	0.244	0.263	0.247	0.246
	Portuguese	0.247	0.252	0.257	0.244
	Spanish	0.246	0.252	0.243	0.259

Table 4.10   Summary of the results for Tables 4.6−4.9. Data presented as mean +/− standard deviation (μ +/− σ). For the Diagonal column, $n = 4$; for the Non-diagonal column, $n = 12$; for the All elements column, $n = 16$. Please see text for further details.

Table (Round trips)	Diagonal (μ +/− σ)	Non-diagonal (μ +/− σ)	All elements (μ +/− σ)
Table 4.6 (0)	1.000 +/− 0.000	0.000 +/− 0.000	0.250 +/− 0.447
Table 4.7 (0.5)	0.868 +/− 0.016	0.044 +/− 0.014	0.250 +/− 0.369
Table 4.8 (1.5)	0.349 +/− 0.008	0.217 +/− 0.034	0.250 +/− 0.065
Table 4.9 (2.5)	0.259 +/− 0.003	0.247 +/− 0.003	0.250 +/− 0.006

for each of the three non-source languages is 74.15%/3 = 24.72%. Thus, only 1.13% of documents are differentially identified, and essentially 99% of all documents are obfuscated. A third additional round trip, involving the final FIPS language, essentially makes all of the elements in the matrix = 0.25 (data not shown). The effect of the round trips (Table 4.6 = 0 round trips; Table 4.7 = 0.5 round trips since it is only translated to English; Table 4.8 = 1.5 round trips; and Table 4.9 = 2.5 round trips) is shown in Table 4.10. As more round trips are added, all matrix elements are asymptotically approaching 0.250, which is the condition of full de-identification of the source language.

This section shows the functional use of translation in obfuscating the source of a document. It should be noted that exactly the same approach could be used to anonymize the source of a document, even if a translation were not needed. In other words, if an anonymous source of information wishes to remain so, their testimony can be "homogenized" using multiple round trips of translation along the lines of the example shown in this section. We next consider a few examples of round trip translation as part of the Section 4.6.

4.6 Test and Configuration

The application of Damerau–Levenshtein distance measurements as a means to assess translations, along with "round trip" translations as a means to "homogenize" translations, are functional means of incorporating translations in your text analytics repertoire. In this section, several translation round trips (using Google Translate, 11 May 2020) are summarized to illustrate these round trips.

The first one is the familiar simple sentence in English that includes each of the 26 characters A to Z: "A quick brown fox jumps over the lazy dog." Here are the results of successive translations between English and Italian:

0.0 Round Trips: A quick brown fox jumps over the lazy dog

0.5 Round Trips: Una veloce volpe marrone salta sopra il cane pigro

1.0 Round Trips: A fast brown fox jumps over the lazy dog

1.5 Round Trips: Una veloce volpe marrone salta sopra il cane pigro

A total of 1.5 round trips in translation (from English to Italian to English Italian again) are shown. No further translation changes will occur since the output of 0.5 and 1.5 round trips are identical. The net result of translation is a synonymic replacement of "quick" with "fast" which means the final English output no longer includes a "Q," an "I," a "C," or a "K" in it, so it is no longer functionally the same kind of sentence. That being said, however, its word-level Damerau–Levenshtein distance is only 1 since "fast" can be replace by "quick" to get the original sentence (a single substitution).

The next example involves translation between English and French, and it also completes its changes by 1.5 round trips:

0.0 Round Trips: I saw him coming but did nothing to stop him

0.5 Round Trips: Je l'ai vu venir mais je n'ai rien fait pour l'arrêter

1.0 Round Trips: I saw it coming but I did nothing to stop it

1.5 Round Trips: Je l'ai vu venir mais je n'ai rien fait pour l'arrêter

Here, the word "him" is replaced by "it" in two locations and an additional occurrence of the word "I" is inserted. The pronoun substitution makes the English sentence feel vaguer, and the Damerau–Levenshtein distance is 3: two substitutions of "him" for "it" and the deletion of the second "I" are required to get back the original sentence.

The final example for this section performs round trips in English and German and involves a difficult sentence for some English speakers to understand. Its translations are frozen by the end of two round trips, as shown here:

> **0.0 Round Trips**: The complex houses married and single soldiers and their families
>
> **0.5 Round Trips**: Der Komplex beherbergt verheiratete und alleinstehende Soldaten und deren Familien
>
> **1.0 Round Trips**: The complex is home to married and single soldiers and their families
>
> **1.5 Round Trips**: Der Komplex beherbergt verheiratete und alleinstehende Soldaten und ihre Familien
>
> **2.0 Round Trips**: The complex is home to married and single soldiers and their families

The expression "houses" is replaced by "is home to" which results in a Damerau−Levenshtein distance of 3 but conveys precisely the same information. Overall, these three examples illustrate how translation engines "converge" on a specific set of expressions, which are often a bit different in meaning than the original, a process we call "homogenization" of the expression. As noted in the previous section, however, this homogenization can provide anonymization for the source of the original content. The homogenized content provides an "acceptable substitution" as a functional replacement for the original text.

4.7 Summary

This chapter describes some of the inherent difficulties of translation, including the differences in how humor and plays on words are comprised in different languages. The need for sufficient training data, related languages, and context to enable accurate translation was discussed. An example of translation gone horribly haywire for a former US President was given, along with some ways to failsafe such systems of translation. Next, a brief review of the approaches to language translation in the literature was overviewed. Our functional approach of using the similarity of behavior between queries, summarizations, and documents as a means of assessing translators was then provided. The weighted percent of query matching for the pairings of (query, summarization) and (query, document), in particular, were described. In the machine learning section, document reading order was described as a functional means of assessing the accuracy of different translators. In the design and system configuration section, we illustrated examples of how to use the translations of multiple dialects to optimize a system for robustness. The correlation between the accuracy of dialect understanding and translation accuracy was shown to be a good predictor of overall translation engine accuracy. The use of multiple, sequential translations was then coupled with measurements of Damerau−Levenshtein distance to predict translation engine accuracy. Functional applications of translation included a study of how to use a translation engine's "native language" as an intermediate translation to improve the overall accuracy of translation between two other languages. Applications of multiple translations to obfuscate the source language of a document and to provide more uniform translations in each language were given. We concluded with several examples of round tripping translations between two languages and with the corresponding "homogenization" of content.

References

[Agar15] Agarwal B, Mittal N, "Sentiment analysis using ConceptNet ontology and context information," Prominent Feature Extraction for Sentiment Analysis, pp. 63-75, Springer, 2015.

[Dame64] Damerau FJ, "A technique for computer detection and correction of spelling errors," Communications of the ACM 7(3), pp. 171-176, 1964.

[DeSi97] De Silva LC, Miyasato T, Nakatsu R, "Facial emotion recognition using multi-modal information," Proceedings of ICICS, DOI 10.1109/ICICS.1997.647126, 1997.

[Emot14] Emotion Markup Language (EmotionML) 1.0, W3C Recommendation 22 May 2014, available on h t t p s : //www.w3.org/TR/emotionml/, accessed 12 June 2020.

[Gues14] Guest post, "The Most Famous Examples of Misinterpretation," A Translator's Thoughts, originally posted 29 April, 2014, https://translatorthoughts.com/2014/04/the-most-famous-examples-of-misinterpretation/, accessed 8 June 2020.

[Koeh10] Koehn P, *Statistical Machine Translation*, Cambridge University Press, New York, 2010.

[Kout15] Koutrika G, Liu L, Simske S, "Generating reading orders over document collections," IEEE 31st International Conference on Data Engineering, DOI 10.1109/ICDE.2015.7113310, pp. 507-518, 2015.

[Lacr18] Lacruz I, Jääskeläinen R, (Eds.), Innovation and expansion in translation process research, John Benjamins Publishing Company, 2018.

[Leve66]   Levenshtein VI, "Binary codes capable of correcting deletions, insertions, and reversals," Soviet Physics Doklady 10(8), pp. 707-710, 1966.

[Liu19]   Liu X, Wang W, "Multiway Attention for Neural Machine Translation." Proceedings of the 2019 2nd International Conference on Algorithms, Computing and Artificial Intelligence (pp. 265-270), 2019.

[Luon15]   Luong MT, Pham H, Manning CD, "Effective approaches to attention-based neural machine translation," arXiv preprint arXiv:1508.04025, 2015.

[Monr17]   Monroe D, "Deep learning takes on translation," Communications of the ACM, https://doi.org/10.1145/3077229, 2017.

[Nagp10]   Nagpal R, Nagpal P, Kaur S, "Hybrid technique for human face emotion detection," IJACSA, vol. 1, no. 6,pp. 91-101, 2010.

[Papi02]   Papineni K, Roukos S, Ward T, Zhu WJ, "BLEU: a method for automatic evaluation of machine translation," Proceedings of the 40th annual meeting of the Association for Computational Linguistics (pp. 311-318), 2002.

[Shan48]   Shannon CE, "A mathematical theory of communication," The Bell system technical journal, 27(3), 379-423, 1948.

[Sims13]   Simske S, "Meta-Algorithmics: Patterns for Robust, Low-Cost, High-Quality Systems", Singapore, IEEE Press and Wiley, 2013.

[Sims14]   Simske SJ, Boyko IM, Koutrika G, "Multi-engine search and language translation," Proceedings of the EDBT/ICDT 2014 Joint Conference.

[Sims19]   Simske S, "Meta-Analytics: Consensus Approaches and System Patterns for Data Analysis," Elsevier, Morgan Kaufmann, Burlington, MA, 2019.

[UNL20]   Universal Networking Language, Wikipedia website, https://en.wikipedia.org/wiki/Universal_Networking_Language, accessed 23 June, 2020.

[Wang20]   Wang X, Kou L, Sugumaran V, Luo X, Zhang H, "Emotion correlation mining through deep learning models on natural language text," IEEE Transactions on Cybernetics PP(99), pp. 1-14, 2020.

[Wu06]   Wu C-H, Chuang Z-J, Lin Y-C, "Emotion recognition from text using semantic labels and separable mixture models," ACM Transactions on Asian Language Information Processing 5(2), pp. 165-183, 2006.

[Yaco03]   Yacoub S, Simske S, Lin X, Burns J, "Recognition of emotions in interactive voice response systems," Eurospeech 2003, pp. 729-732, 2003.

[Yan02] Yan Q, Vaseghi S, "A comparative analysis of UK and US English accents in recognition and synthesis," IEEE ICASSP, I-413 – I-416, 2002.

[Youn00] Younger J, "Linear A texts in phonetic transcription: 7b. The Script," University of Kansas, 2000.

5

Optimization

"Premature optimization is the root of all evil"
– Donald Knuth

"We minimize risk while we maximize activity. It's this constant balancing act that we do"
– Hope Jahren

"If I had more time I would have written you a shorter letter"
– Blaise Pascal

Abstract

Optimization is shown to be context-dependent and, thus, a functional setting of a text analytical system. We consider optimization in light of its input, its operations, and its output. Input can be readily changed through different levels of compression, or summarization, of the content. Operations to be considered for optimization include those that vary by time, location, language, and adjacent uses and applications. Outputs are the impact that the optimized system has on its environment – its users, its downstream algorithms, and various measurements of its overall acceptability. Thus, optimization is shown to be a process, rather than an end point, and as such is shown to be a functional text analytic. This alternative view to optimization is applied to important functional text analytical systems.

5.1 Introduction

Understanding of the word "optimization" is not optimized. Optimization may well be the "spirit" of every machine learning (ML) project because it can be what drives a lot of the processes. For many ML professionals, ourselves included, there is an almost mystical belief at some point in their career that with just the right input, and just the right relationship between

169

the different features of the input, there is a miraculous convergence on a system that performs so much better than it did before. But, like a spirit, this is intangible, or at least ephemeral, blown to the four winds when the next round of input data is trialed against the system. Optimization is transitory, transitional, and transformational. Optimization is a process, not an accomplishment.

Why is this so? Why is optimization not a setting that can be dialed in? At least partly because optimization is context-dependent; that is, it is a function of the particular conditions under which it is computed. It is a short-term, though still functional, setting of a text analytical system. When Don Knuth – with his eyes blind to greed, sloth, narcissism, envy, and rancor, evidently – states that "premature optimization is the root of all evil," he is by his own account 97% right. This figure comes from Knuth noting that in a critical 3% of situations, we actually should perform optimization. We here argue against that 3% because, in our experience, optimization is a function of its input, its operations, and its output. In a linear system, that is pretty much everything. Therefore, all optimized systems are conditional. If we think the conditions are representative of a good portion of the expected conditions for the system to be exposed to thereafter, then the optimization is likely relatively resilient.

The conditionality of optimization will be shown here with a simple, throw-away experiment, which uses just 100 text documents belonging to four different classes and uses two text classification engines to classify the 25 documents/class. If all 100 documents are correctly assigned, the accuracy = 1.000. If 76 of the documents are correctly assigned, the accuracy = 0.760. The details of the experiment are less important than the fact that it gives us a chance to show how different considerations of the input to a system change the optimization approach.

Input can be readily changed through different levels of compression, or summarization, of the content. Two examples of this are shown in Figure 5.1, in which there are two different text classification engines being used to perform classification on an original set of documents (percent compression = 0.000) or documents compressed by 10%, 20%, 30%, 40%, 50%, 60%, 70%, 75%, 80%, 85%, 90%, 95%, or 97.5%. These compression levels correspond to summaries comprising 90%, 80%, 70%, ..., 10%, 5%, and 2.5% of the documents. Looking at the two trend lines, it is clear that Engine 1 provides better overall accuracy across the entire range of compression values, with a mean accuracy of 85.2%. The mean accuracy of Engine 2 is 82.5%, or 2.7% less than Engine 1. These are validation data being used to recommend the settings for the two engines in deployment. If we optimize at the engine level, wherein we do not know what level of compression we will have for the documents in the actual deployed system, we would select Engine 1 (at 0.852 accuracy) over Engine 2 (at 0.825 accuracy). However, one form

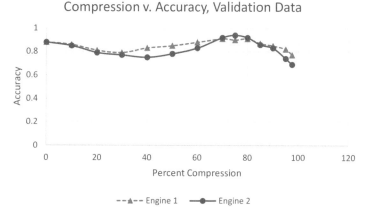

Figure 5.1 Plot of percent compression (amount of reduction in document size during summarization), *x*-axis, against the accuracy of document classification (*y*-axis) for two different document classification engines. The mean accuracy of Engine 1 is 0.852, while that of Engine 2 is 0.825. The peak accuracy of Engine 1 is 0.91 for both 70% and 80% compression, while the peak accuracy of Engine 2 is 0.94 for 75% compression.

of optimization would state that we use Engine 2 with 75% compression (summarization at 25% the original length of the documents) because this provides our highest overall accuracy.

For the validation data (shown in Figure 5.1), the peak accuracy of Engine 1 is 0.91 when employing either 70% or 80% compression, while the peak accuracy of Engine 2 is 0.94 when employing 75% compression. If we optimize our system based on the best overall results on any of the 28 validation settings (either engine with one of the 14 settings for compression), we would select Engine 2 with 75% compression. If we anticipate a range of compression between 70% and 80% for the documents, then the validation data for Engine 1 predicts 0.907 accuracy and the validation data for Engine 2 predicts 0.927 accuracy since those are the mean values of accuracy for these two engines over that range. Given these findings, we can design at least four different strategies for using the input (that is, validation) data to perform system optimization. They are as follows.

(1) Use the optimum setting from among the 28 measurements. This amounts to "Use Engine 2, 75% Compression" for the strategy. That is, we anticipate that all documents will be compressed by 75% and that we will use Engine 2 for the analysis.
(2) Use the optimum setting from the optimum 10% range. This amounts to "Use Engine 2, 70%−80% Compression" for the strategy. That is, we

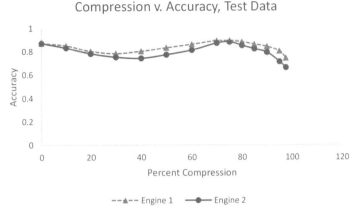

Figure 5.2 Plot of percent compression (amount of reduction in document size during summarization), *x*-axis, against the accuracy of document classification (*y*-axis) for two different document classification engines. The mean accuracy of Engine 1 is 0.835, while that of Engine 2 is 0.795. The peak accuracy of Engine 1 is 0.89 for both 70% and 75% compression, while the peak accuracy of Engine 2 is 0.88 for 75% compression.

anticipate that all documents will be compressed by between 70% and 80% and that we will use Engine 2 for the analysis.

(3) Use the optimum of the two engines, irrespective of compression, which can vary from 0% to 97.5%. This amounts to "Use Engine 1, Variable Compression" for the strategy. That is, we anticipate that the documents will be compressed by some random amount between 0% and 97.5% and that we will use Engine 1 for the analysis.

(4) Use the optimum of the two Engines, irrespective of compression, which can vary from 0% to 97.5%., but then set the compression level to be the optimum amount determined for the non-optimum engine (that is, Engine 2, for which it is 75%). This amounts to "Use Engine 1, 75% Compression" for the strategy. That is, we anticipate that all documents will be compressed by 75% and that we will use Engine 1 for the analysis.

In order to test these four strategies, we must first measure the accuracies of both engines at each level of compression for the test data. The test set comprises another different set of 100 documents, which are 25 each from the same four classes as the input, or validation, documents. The results for the test data are shown in Figure 5.2.

The results for the test data (Figure 5.2) look similar to that of the validation data, with slightly lower accuracy across the range of compression. Here, the mean accuracy of Engine 1 is 0.835 (a drop from 0.852 on the validation data), while that of Engine 2 is 0.795 (a drop from 0.825 on the validation data). The peak accuracy of Engine 1 is 0.89 for both 70% and 75% compression (below the peak of 0.91 on the validation data) and the peak accuracy of Engine 2 is 0.88 for 75% compression (well below the peak of 0.94 on the validation data). We now explain what the results were for each of the four optimization strategies (this is summarized in Table 5.1, as well).

(1) If all the documents in the test set are compressed by 75% and that we used Engine 2 for the analysis, we obtained 0.88 accuracy. This is 0.01 lower accuracy than using Engine 1 on the 75% compressed test documents. In Table 5.1, this results in a "Δ (delta) from Peak Accuracy" of 0.01.

(2) If the test documents are compressed by 70% ($n = 100$), 75% ($n = 100$), and 80% ($n = 100$) and we use Engine 2 for the analysis, we obtained 0.867 accuracy. This is 0.02 lower accuracy than using Engine 1 on the 70%−80% compressed test documents. In Table 5.1, this results in a "Δ from Peak Accuracy" of 0.02.

(3) Using Engine 1, with the 14 compression levels being equally adopted. For these, we compressed the 100 documents at 14 different compression levels, as described above, ranging from 0% to 97.5%, and pooled all of the results. Using Engine 1 on this array of compressions, we obtained an accuracy of 0.835, which is the peak accuracy since the accuracy for Engine 2 was only 0.795. This strategy is nowhere near as accurate as (1) or (2) since the documents compressed less than 70% or more than 80% had lower accuracies than those in the 70%−80% compressed range.

(4) The fourth scenario used the better performing engine, Engine 1, but with the documents compressed only at the level for which the other engine, Engine 2, had peak accuracy (that is, at 75%). This resulted in the highest accuracy for the testing data; that is, 0.890.

The results of these four strategies being applied to the test data are summarized in Table 5.1. The single best overall optimization strategy is Strategy 4, which uses the optimum setting from the validation data for Engine 2, applied to the use of the engine with more consistent performance across the entire range of compressions, Engine 1.

Why does Strategy 4 in Table 5.1 provide the highest overall accuracy? Well, for one thing, we should not read too much into this result since it was performed on only 200 documents total (even if, in the case of Strategy 3, each document was compressed by 14 different levels). The example is simply given to illustrate how many different "optimizations" can be claimed for even such a simple experiment as this. We can optimize by validating

Table 5.1 List of different strategies for optimization and their corresponding test accuracy (accuracy on the test data in Figure 5.2), the peak accuracy for this strategy (higher of the accuracies of Engine 1 and Engine 2 for the strategy), and the delta () between the two, where $\Delta = 0.000$ means the strategy for optimization was the most accurate choice.

Strategy	Test accuracy	Peak accuracy	Δ from peak accuracy
(1) Use Engine 2, 75% Compression	0.880	0.890	0.010
(2) Use Engine 2, 70-80% Compression	0.867	0.887	0.020
(3) Use Engine 1, Variable Compression	0.835	0.835	0.000
(4) Use Engine 1, 75% Compression	0.890	0.890	0.000

the results against specific settings for each of the specific engines, and the settings that give us the best results for the validation data are assumed to be a global, or at least near-global, optimum. That is the case for Strategy 1. Or, we can find a neighborhood around this optimum and use that neighborhood to select a presumably more resilient optimization that will not change as much to changing input. With the small number of documents in this experiment, that was not the case, and Engine 2 achieved its highest accuracy for 75% compression in both validation and test sets. However, Strategy 2 is a reasonable approach to larger data sets since generally the global optimum point will have some variability with changing input. The third strategy extended this "resiliency" to cover the entire range of input. Clearly, this was overly ambitious and effectively assumes that there is no optimum compression range and that accuracy will be maintained across the range. The last strategy may have worked in this case – and shows promise of working well in other cases – due to incorporating aspects of the optimization of both engines on their validation data (optimal full-range accuracy of Engine 1 and optimal compression settings of Engine 2). Regardless, the point is driven home here: for a given set of inputs, there is a plurality of different interpretations for what "optimization" is.

Operations of tunable systems can also be optimized. Operations are a broad set of capabilities, including the collection of system-relevant data and its interpretation (analytics), transformational approaches to improve the value of the data (algorithmics), and associating the data with other content due to similarity, correlation, or other adjacency (re-purposing). Operations include a wide range of attributes – such as time-stamps, location information (including GPS), language used, creator, sender, intended recipient, purpose, and a host of other metadata – that provide cues as to how to use, re-use, and re-purpose the content. A time-stamp of "2011" on traffic data, for instance, means something quite different in Youngstown, Ohio, than it does in Bend, Oregon since one city has substantially less traffic presently, and the other considerably more. Without such attribute tagging, data is just data, unstructured, contradicting, and, in its extreme, not helpful. With the attribute tagging, however, we can significantly upgrade the value of the data. Hope Jahren said it in another, more serious, context, but the comment that "we can minimize risk while we maximize activity" is as good a statement about optimization as any I could find. The risk that we offset when we maximize the activity of our data is the risk of potential loss of this incremental value being added. Hope Jahren is also correct when she notes that this is a balancing act. The most difficult balancing act is on a knife's edge, which is often the case for the overall global "optimum" − think of the balance beam in gymnastics. Instead, it is far easier to balance on a wide flat plane − think of the floor for your yoga class. No doubt someone who is balancing on a beam is an optimized athlete, but the floor of a yoga studio is optimum for more people. It all depends on what "optimum" means.

The area (or volume, or higher-dimensional volume for spaces of four or more dimensions) around the optimal point can be used for one useful "optimized" definition of an optimum. We call this region around the given optimum its neighborhood and its measurement as area, volume, etc., its size. Given this, we can define an "optimal" optimum as given in Equation (5.1):

$$\text{Optimum}_{\text{Optimal}} = \max_{\text{optima}} \left((\text{optimum} - \mu) \left(\text{Size}_{\text{Neighborhood}} \right) \right). \quad (5.1)$$

Here, $\text{Optimum}_{\text{Optimal}}$ is the overall "best" optimum as defined by that whose optimum value as a difference from the mean value of the entire output space, or (optimum-μ), multiplied by the size of its converging neighborhood, has the highest value. So, suppose that for a reasonable range of settings, we have several identified optima. We then determine the neighborhood around each optima that converges to the same optimum using any of the wide variety of optimization approaches (evolutionary algorithms, dynamic programming, Newton's method, nearest neighbor search, etc.), and tabulate it. An example of this is shown in Figure 5.3.

In Figure 5.3, we assessed several hundred randomly distributed points in the space and then determined points around each maximum that converged to the maximum using, in this case, Newton's method. Four non-trivial optimums with neighborhoods were identified and labeled {A, B, C, D}. The peak value in each neighborhood along with the size of each neighborhood was computed. The mean value of the space around the optima neighborhoods was 80, meaning the peaks of {A, B, C, D} are {65, 100, 290, 170} above the mean value. Using Equation (5.1), these relative peaks are then multiplied by the sizes of the neighborhoods to yield values of {3250, 3000, 2900, 4250}. Here we see that we have selected neighborhood D in spite of the fact that the global optimum is within the neighborhood C. This is because the larger the neighborhood, the more likely the optimum therein

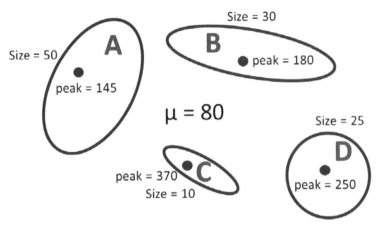

Figure 5.3   Plot of four optima in a space for which the mean value is 80. The relative weight of each optimum {A, B, C, D} is, as per Equation (5.1), Weight$_A$ = (145 − 80) × 50 = 3250; Weight$_B$ = (180 − 80) × 30 = 3000; Weight$_C$ = (370 − 80) × 10 = 2900; and Weight$_D$ = (250 − 80) × 25 = 4250. We thus select "D" as the optimal optimum, even though its peak of 250 is below that of "C," which is 370.

is to be stable and resilient to changes in the input. However, the magnitude of the optimum is also important for accuracy; so the peak value (above the mean) is multiplied by the size to provide a system at least theoretically optimized for both stability and accuracy.

Other, more complicated, means of selecting an "optimal optimum" are, of course, possible. Feeding back some information from the outputs of the system allows the system designed to take into account the impact that the optimized system has on its environment. One of the simplest means of incorporating feedback is with a Likert scale [Like32], which has been around for a long time. Here, users can give "constrained" quantitative feedback on the performance of the system under different optimization strategies. A more functional method – which is therefore more closely in keeping with the themes of this book – is to collect use-based analytics of downstream algorithms to evaluate the effectiveness of each optimum. Let us imagine that, in the example above, {A, B, C, D} are different filtering approaches to voice recordings (that is, human speech), corresponding to different combinations of low-pass filtering, notch (60 Hz) filtering, high-pass filtering, and the like. We find four different optimum sets of filters that each does a much better job of filtering the speech for human understanding than other settings near them. So, these four settings can be functionally compared by one or more of the following approaches:

(1) People listen to the filtered speech and give it a quality rating on a Likert scale (e.g., "rate the quality from 1 to 10, 10 being the highest"). While simple enough to implement (think of "Google Forms" as one mechanism), the downside of this approach is that it requires human ground truthing, which is not the preferred method when a more functional method can be accommodated.

(2) We place the speech in each of the four filtered forms in an automated call center and collect analytics on listener response to the four differently filtered expressions. Analytics include latency of listener response, the number of confusion-indicating responses (especially questions like "What did you say?" or "I don't understand you"), and indirect indicators of confusion, such as anger [Yaco03].

(3) People in the call center ask for a human operator more quickly with one filtering selection versus the other.

(4) The filtering approaches are applied to speech in one language which is then translated into another. The second language speech is then evaluated automatically for spoken language proficiency in the second language [Gret19].

Of these four approaches, we prefer (4) since it can be completely automated, requiring no human ground truthing. It is also a completely functional text analysis since it is using language proficiency as a measure for the best filtering approach. As in many of the other functional approaches in this book, it illustrates how a given system optimization problem is assessed by a functional measurement of its overall acceptability through incorporating it into a different task. In this example, then, optimization is shown to be a *process*, rather than an *end point*.

We arrive at the conclusion that optimization is conditional and, thus, adaptable. There is a lot of flexibility in this: optimization can be tied to absolute (e.g., accuracy) or relative (does selecting optimization differently result in improved accuracy?) objective functions. We have already demonstrated (Figures 5.1 and 5.2) how to optimize a summarization strategy, along with its percent compression. As we have established, sometimes, summarized versions of documents function better than the original document. This is hardly surprising since we are already very familiar with the fact that removing stop words ("the," "which," "is," "or," etc.) [Luhn59] usually improves the functional use of a document when treated as a vector of word occurrences. As Blaise Pascal (a none-too-shabby statistician in his own right) might have had in mind when he said, "if I had more time I would have written you a shorter letter," the ability to simply vary the amount of compression in a summary gives us an important independent variable for tuning a functional text analytic. In addition, virtually, any other text processing task can be performed using summarized representations of the documents. While part of speech tagging and word sense analytics are usually expected to perform better on larger text data sets, it is sometimes possible to get better performance on the difficult-to-analyze portions of a document corpus by employing compressed text content (that is, summaries). This seems contradictory, but the fact that only the most salient text is analyzed means that the natural language processing (NLP) algorithms might identify unusual word usage, including non-traditional word sense, better than if the less salient portions of the document "overwhelm" the analytics. That is, slang, jargon, and other neologisms are concentrated in the summaries in comparison to the document as a whole and so provide a higher density for these linguistic "anomalies" than does the entire document. A summary can thus be a "focused challenge" to a text analytics system. It is, therefore, unsurprising that functional approaches to optimization can readily be performed for each of the following text analytics tasks:

(1) Clustering
(2) Classification
(3) Categorization
(4) Tagging/indexing
(5) Query set generation
(6) Translation

The first three were the focus of Chapter 3, but it is worth noting here that the assignment to a cluster, a class, or a category can be assessed (that is, functionally compared), analogous to the speech filtering example above, using one or more of the following approaches:

(1) People look at the top-ranked cluster, class, or category assignments and give them a quality rating on a Likert scale (e.g., "rate the relevance of the assigned cluster from 1 to 10, 10 being the highest"). The highest overall rated assignment is accepted. Again, this type of approach has the downside of requiring human ground truthing, certainly not the preferred method if it can be avoided.
(2) We place the top-ranked assignments into a data mining algorithm that is meant to extract specific data fields from properly assigned documents. For example, if the document has been classified as a bank loan application, fields to be extracted include "applicant," "home address," and "e-mail address." If proper fields cannot be extricated, then it is likely that the document has been misassigned.
(3) People performing searches using queries with one or more of the cluster, class, or category labels select these documents when they are returned with the search. This is use-based, functional verification of the proper assignment of the clusters.

Tagging, indexing, and query set generation are the complement of clustering, classification, and categorization, as they represent the manner in which sets of documents are assigned specific descriptors, often the words with the highest TF*IDF (term frequency times inverse document frequency) values in the document sets. Translation, of course, has many functional optimization possibilities, some of which were overviewed in Chapter 4. With this relatively long introduction to optimization, we now look at the general considerations as they apply to the broader functional text analytics approaches.

5.2 General Considerations

The first attribute of optimization to understand is that optimization is iterative. Except for exhaustive search, in which every possible outcome of a system is evaluated and graded, every other text analytics optimization task involves an initial estimate for the optimum system settings, followed by iterative refinement of the estimate until changes between one iteration and the next are either smaller than some small difference or repetitive (e.g., cycling between final states F_1 and F_2); alternatively, the number of iterations has exceeded some maximally allowable value, and the system accepts whatever the output is at the end of the allowed number of iterations.

How can functional text measurements be used to iteratively improve a system? The simple answer is no differently than any other objective function. It may seem a little odd to have a functional metric in an area only related to the actual output that we wish to optimize. However, as was shown in the example above for the four strategies for optimization (Table 5.1), the use of a functional metric to optimize the output of a different, but related, metric is a good path forward for overall system resiliency. Another aspect of pairing two different metrics with at least somewhat distinct functional goals in mind is the fact that this comes closer to exploring a wider gamut of the system's overall application space. Thus, it is more comprehensive in addition to being more resilient. Also, feeding information from one application space to the other works some magic of "feedback" in other control systems. It provides in some ways a "feedback control" for the overall text analytics.

The pairing of two or more methods can be considered a "meta-method," and we have seen several different approaches to that in previous chapters. In clustering, a classic algorithm for trading off two approaches is expectation–maximization (E-M), which consists of the following steps [Demp77]:

(1) Provide an initial estimate of the relevant output of the system. In the case of using E-M to find an optimal set of clusters for a set of data,

the centroids of the k clusters are estimated. This can be performed randomly; semi-randomly (wherein N random sets of initial centroids are generated, and the one with the lowest sum squared distance from all of the samples is chosen as the starting point); or with some sort of expert system (e.g., historical centroids for these types of clusters) input. This is the "E" in "E-M" and stands for the "Expected" distribution (even if the initial expected distribution is not very accurate).

(2) Collected input data is fed into the system. These may be initially the same data used to estimate the "E" in (1) if anything other than a random assignment was used.

(3) The statistics about each of the initial k sets of data (that is, their probability distributions) are updated with the data. In the simplest form, that of performing clustering, every data point is assigned to the centroid nearest it. The centroid is then updated to maximize its fit to all the data assigned to it. This maximized fit is the "M" in "E-M."

(4) These three steps, (1)–(3), are repeated until the output is stable (it does not change, change a trivial amount, or oscillate between two solutions).

Other iterative methods for data assignment processes such as clustering include affinity propagation [Frey07] and the Markov chain Monte Carlo [Geye92] method. We also described the competition–cooperation method in Chapter 1: this approach uses a competitive algorithm in the first half of an iteration and then uses a cooperative algorithm in the second half of the cycle. One form of competition–cooperation is the use of a Tessellation and Recombination approach [Sims13], which can also be applied to aggregation of data. Here, the results of multiple analytics are combined to break up a data set into as many subclusters as possible ("competition" here is the competition of the algorithms' different representations of the output). Then, clusters are recombined using a specific algorithm for combination. One such algorithm is to iteratively combine two clusters (out of the N clusters currently existing) such that the increase in the mean squared error (distance) of the overall set of data is minimally increased when the new cluster centroid is formed and the number of clusters changes from N to $N - 1$. This requires $N(N - 1)/2$ different cluster combinations (the "cooperation" step in the iteration), but it is rather effective in creating new clusters that are meaningful.

This process can be terminated in a large number of ways. One straightforward means of stopping it is to evaluate the percent change in sum squared error when performing optimum clustering (combining) of two clusters. An example is shown in Table 5.2, in which 10 clusters are sequentially reduced to 4 clusters. When moving from 10 to 9 to 8 to 7 clusters, the sum squared errors increase by 3.2%, 7.6%, and 10.8%.

Table 5.2 Example of using the Tessellation and Recombination for clustering. The clustering algorithm finds the merging of two existing clusters that cause the smallest increase in sum squared error of the elements in each cluster from the centroid of the combined cluster, and chooses to merge those two. The percent increase in overall sum squared error (last column) indicates clustering should stop at N (Clusters) = 7 since the low percentages leading from 10 to 7 clusters (3.2%, 7.6%, and 10.8%) rise considerably when trying to move to 6 or less clusters (31.5%, 33.7%, and 42.2%).

Iteration	N (Clusters)	Sum squared error	Change in sum squared error
1	10	14,532.0	N/A
2	9	14,994.7	+3.18%
3	8	16,132.5	+7.59%
4	7	17,867.7	+10.76%
5	6	23,497.6	+31.51%
6	5	31,425.7	+33.74%
7	4	44,674.2	+42.16%

However, when further combining clusters to move from 7 to 6 to 5 to 4 clusters, the sum squared errors increase much more; that is, by 31.5%, 33.7%, and 42.2%. It is clear from these values that no less than seven clusters should be formed.

Taken more broadly than at the simple algorithm level, optimization is a process by which an existing solution is re-evaluated with an emphasis on change. Our goal in optimization is to sufficiently explore the input domain such that no suitably large products of optimum magnitude and the size of its neighborhood of convergence (Equation (5.1)) are missed in our search. Traditionally, evolutionary algorithms and Monte Carlo approaches are reliable for exploring the input space; however, other methods (like artificial neural networks or ANNs) are usually a lot faster in the deployment phase of the project. However, ANNs generally require a lot more training data for them to provide a system resilient to large-scale changes in input. This is often the case when ANNs have many extra edges (connections between layers in the net) than they need to cover all of the training data, and without the proper learning algorithms do not distribute the weight in an entropic fashion across all of the edges. Instead, they can use a few connections at each level to handle the learning. However, ANNs are increasingly capable of relatively straightforward self-training, which nicely suits the argument for functional means of generating text analytics. Along with the other optimization methods described so far in this chapter, then, ANNs avoid the cost, uncertainty, and delay associated with bringing human ground truthing into the optimization problem; for example, using Amazon Mechanical Turk has uncertainty about the credentials, motivation, and reliability of the humans involved.

In these two sections, we have shown optimization to be context-dependent. It is also intention-dependent; that is, a function of what the system designers are hoping to get out of the system. If the system is meant to be resilient to large-scale changes in the input – including such considerations as changes in amplification, throughput, and periodicity – then our optimal optimum is one that will be most resilient to such changes. This argues against a system in which training, validation, and test sets are highly similar, which is the usual case for reporting the accuracy, performance, and reliability of most analytics systems, not just those being used for text analytics. Instead, we may wish to create asymmetric content in these sets, swapping them out and then creating the settings for the deployed system as a combination of these multiple experiments.

We also saw that sensitivity analysis of a sort can be performed on the input by changing the level of compression. For text analytics, intelligent compression can be performed in at least two ways. The first is what we showed in the example of the previous section; that is, summarization. However, it should be pointed out that summarization is not necessarily the optimum manner of compressing a document for re-purposing. A summary, unless it is unreadable, will include a large proportion of stop words and other common words (like prepositions, articles, common nouns, and commonly used adverbs, verbs, and adjectives) that do not differentiate it from other documents. Thus, a second way of compressing a document is to reduce it to its key words (also referred to as indexed words, tags, etc.) and use these to represent the document in its functional task. One relevant set of keywords for a document are the words which have the highest TF*IDF values when comparing the frequency of the words in the document (term frequency) to all the other documents (document frequency) of relevance in the document corpus. Such a keyword-based approach to compression is a form of optimization in which for each percentage of compression, the remaining words representing the document are, in theory, the set of words that best distinguish it from other documents.

In addition to the input, the actual operations of the text analytics can be optimized. Optimizing text analytics for time, for example, might require the conversion of the text into a time-series for each character. The distribution of terms in the document might be as meaningful as the actual word counts. For example, two words occurring with the same frequency in the document overall, but with one concentrated in a single section of the document and the other distributed relatively uniformly throughout the document, likely play different roles in the meaning of the document. Thus, an optimal representation of the role of a word in the document includes at minimum: (a) its absolute frequency (word count); (b) its relative frequency (TF*IDF);

and its distribution (document entropy). Which of these is optimal, again, depends on the context of the situation in which it is to be used.

The output of text analytics covers a large gamut. Text analytics are so pervasive that they are often taken for granted. Unfamiliar words, such as slang or neologisms, stand out even in a language with well over 100,000 words in it. Humans are, in general, good detectors of anomalies, and, perhaps, nowhere is this as obvious as in language. Subtle differences in accent are immediately associated with different regions. The pronunciation of a single word (for example, the word "about" as pronounced by a man from Manitoba versus a boy from Boise) can give away nationality, regionality, all the way down to a district in a city (Cockney, Boston Southies). Accent, dialect, and regional languages can be optimally identified, just like documents, by the pronunciations and expressions that most stand out from all of the other accents, dialects, and regional languages. The output of speech can have a tremendous impact on its environment: either saying "Fuhgeddaboudit" or pronouncing your "th" like a "d" will identify yourself as a New York City metro person to a woman in Lubbock, while saying "howdy" or pronouncing "well" like "wail" might do the opposite. One of the most surprising facts about accents is that you have one. Speaking English as my first language, it is easy to forget that when I order a coffee in Calgary, fish and chips in Chichester, or pasta in Perth, it is my English that has an accent. My speech is optimized to my environment, and so the clarity of my utterances is, like everything else in language, conditional. The "downstream algorithms" for my speech are the abilities of other people to convert my cepstrals into something sensible. And, simply talking louder is not going to make it easier, unless in speaking louder, I somehow take on a local accent better. However, after a week in Winnipeg, a month in Melbourne, or a fortnight in Falkirk, I will find myself amazed at how well those folks have adapted to my accent! In reality, my pronunciation has been slowly optimized for its purpose; that is, of conveying my thoughts to those around me. Speech is therefore just another form of functional optimization.

5.3 Machine Learning Aspects

ML algorithms for speech, as for any other area of ML application, are highly dependent on the **quantity**, **quality**, **breadth**, and **depth** of training data. **Quantity** is important in language because the vocabularies of modern languages are immense, and, every year, hundreds of new words are added simply because of the advance of technology, science, international communication, and invention. Quantity also means several things for ML and language. The working vocabulary of a few thousand words that are used in every day speech is almost universally important for text analytical tasks. However, depending on the context, training data will have different quantitative needs for the rest of the language's complete vocabulary. A biomedical engineer is expecting to see terms like biocompatibility, electrical isolation, and fatigue testing, while a psychologist anticipates behaviorism, rationalization, and schizophrenia to terms readily handled by whatever text analytical system is being used. The **quality** of the content is also important. A term such as "biocompatibility," for example, will ideally be associated with some of its key contextual space in the training set, including its governing standards, its associated tests and measurements, and its timing in product life cycle for developing new subcutaneous implants, among others. One of the typical shortfalls of training data is that the wide variety of contexts in which a word or expression is used are hard to validate in the training set. This is a bit of an Ouroboros since, of course, linguistics can help us understand these contexts, but the overall utility of the linguistics derives from being trained on such a balanced set of context.

Breadth and depth of training data are also of high importance. **Breadth** is, of course, related to the quality of the content, which is heavily impacted by the comprehensiveness of the contextual situations for the training data. However, breadth is also associated with training data that is pre-adapted to data drift such as changing accents (which allows one to read Shakespeare without difficulty) and changing meaning in expressions over time. Words that sound like racial slurs in spite of having very different etymologies, for example, are less condoned during certain times of history than in others. Words like "irregardless" and "ain't" are certainly more acceptable now than in the past. Breadth of training sets includes breadth by time (language use over the years) and association (occurring in close context with other words). **Depth** of training data is also very important and largely consists of having enough examples of the input to represent its frequency and its usage.

An important part of all ML designs is the ability to incorporate user feedback on the output in order to improve the system accuracy, performance, robustness, costs, and other factors. A well-designed ML-based system requires only changes in the settings in response to this feedback, not changes in the overall design. This means that the feedback should be as closely tied to the settings as is reasonable. If, for example, the ML system has a setting based on the comprehensibility of the output (with lower comprehensibility using a more core vocabulary, and higher comprehensibility using a more extensive vocabulary), the setting for the user feedback might be on comprehensibility in an A/B testing scenario ("which of these two is more comprehensible?") in a Likert scale ("on a scale of 1–10, rate the comprehensibility of the text, with 10 being the most comprehensible").

From simple feedback such as this, text differences can be very quickly assessed, in some cases simply by substituting one word or expression for another. This allows us to model the text as a mathematical function with the changes in words or expression being the partial derivatives of this function. Consider the sentence S:

> *The man truculently denied operating the vehicle under the influence.*

Now, if we take the partial derivative of S with respect to truculent – that is, $\partial S/\partial$(truculent) – and set it to be greater than zero so that $\partial S/\partial$(truculent) > 0, we might get:

> *The man belligerently denied operating the vehicle under the influence.*

We can also set the partial derivative with respect to the expression "under the influence" to be greater than zero so that $\partial S/\partial$(under the influence) > 0, and get:

> *The man truculently denied operating the vehicle inebriated.*

In each case, we can investigate the impact of this change on the positivity of the response of users to the information (marketing application) or the value of the so-altered information in functional text analytics such as summarization, translation, clustering, and classification. Thus, a major concern for ML is whether or not small changes in content can affect the text analytics output. In some cases, mapping all synonymic terms to a single term certainly helps (e.g., for indexing, keyword generation, and translation); in others, it may not be as useful or even detrimental. In the latter case, a good example is in summarization, where mapping all of the synonyms to the same term makes the mapped-to term more common and thus often less likely to be represented in the summary.

5.4 Design/System Considerations

The more complicated a system becomes the more options there are for optimization. This means that optimization becomes both a harder task and a task that invites creativity at the same time. One of the most common ways for an optimization problem to quickly become intractable for "exhaustive search" approaches is when the text analytics expert is trying to sequence several text analytics processes, or steps, at a time. Multiple steps in a text analytic process are referred to as a "workflow" since these steps are considered progressive, in as much as they upgrade content after each step.

In a workflow, the number of choices at each step is multiplied together to compute the total number of possible workflows. However, sometimes, choices at a current step are incompatible with some of the choices made in the previous steps. This can greatly reduce the number of different workflow combinations that need to be tested. Perhaps, the simplest example of a workflow in which the total possible sequences can be significantly reduced is when one or more steps in the pipeline (the name for a "workflow sequence") must occur in a limited set of languages, while other steps can occur in many more languages. Suppose that we have the following pipeline, with the number of possibilities at each step indicated:

(1) Image capture of text in the environment − e.g., billboards, street signs, menus on restaurant doors, etc. There are hundreds of different capture devices in the world, including digital cameras, mobile phone cameras, and wearable cameras. Among digital cameras, there are four different classes: bridge, compact, DSLR, and mirrorless. Using just the three major phone platforms, these four digital camera types and surveillance cameras gives us eight different input device classes.

(2) Optical character recognition (OCR) of 202 languages is possible [Abby20], and so we have the ability to set up 202 pipelines for the technology to extricate the text from the images and convert them into digital text strings just like the characters in this sentence.

(3) Word histograms of the OCR output are computed. Three are salient: (a) removal of stop words followed by stemming; (b) removal of stop words followed by lemmatization; and (c) representation of the words in TF*IDF representation.

(4) Summarization of the text. Since there are 202 languages supported, there are 202 summarizers to be developed (e.g., we have chosen not to use Text Cloner Pro as part of the OCR engine since we want more control over the summarization).

(5) Prepare the output of the summarization for search, indexing, and classification and perform the selected task. There are three potential options here.

From the above five steps, which ingest a picture of text in the environment into labeled text, we have a maximum of $8 \times 202 \times 3 \times 202 \times 3$ pipelines, if each step is independent of the others. This is a total of 2,937,888 pipelines. However, suppose that we have a highly advanced summarization engine that performs only in English, and like the example in Chapter 4, we have a translation engine developed with English as the core language. We then choose the following pipelines:

(1) Image capture of text in the environment: eight choices

(2) OCR of 202 languages

(3) Word histograms of the OCR output: three choices
(4) Translate into English (if needed)
(5) Summarization of the text in English
(6) Translate back from English (if needed)
(7) Search, indexing, and classification: three options

This reduces the total number of possible pipelines to 14,544. Importantly, 202 parallel pipelines are what are actually developed. This means that there are actually only 72 core pipelines choices (image capture device, word histogram preparation, and output choices multiplied together) to be optimized, assuming they do not affect the OCR differently from one language to another (which needs to be confirmed for at least three widely different languages before asserting).

Another important system issue we need to consider at the onset is the intended breadth of the system. This is, perhaps subtly, different from system resilience, at least as we view it. System resilience is a measure of the variance in the system based on the variance in the input. We can measure the instantaneous change in resilience by making a single change in the input and the compute the system resilience from the ratio of the change in input divided by the change in accuracy. The smaller the change in accuracy to a change in input, the more resilient the system. We define this as shown in Equation (5.2).

$$\text{Resilience}_{\text{System}} = \frac{\partial \text{Input}}{\partial \text{Accuracy}}. \tag{5.2}$$

Here, ∂Accuracy is given by Equation (5.3):

$$\partial \text{Accuracy} = \text{Accuracy}_{\text{new}} - \text{Accuracy}_{\text{old}}. \tag{5.3}$$

Let us illustrate this by using the example from above of our truculent, possibly imbibing vehicle driver. The original sentence reads:

The man truculently denied operating the vehicle under the influence [Sentence 1]

Next, we change the sentence to:

The man belligerently denied operating the vehicle under the influence [Sentence 2]

We have substituted "truculently" in Sentence 1 with "belligerently" in Sentence 2. Using a semantic similarity calculator, "belligerently" is a word with 0.9 similarity to "truculently." Ignoring the articles "the," there are seven words in these sentences, and the original sentence has a 7.0 (perfect) match to itself, while the second sentence has 6.9/7.0 matching, as shown here:

Man (1.0) belligerently → truculently (0.9) denied (1.0) operating (1.0) vehicle (1.0) under (1.0) influence (1.0) [Sentence 2 → Sentence 1]

We then calculate the percent change as the difference between the two sentences, divided by the number of terms in the original sentence, as shown in Equation (5.4).

$$\partial \text{Input} = \frac{|\text{Terms in Original Sentence} - \text{Similarity}|}{\text{Terms in Original Sentence}}. \tag{5.4}$$

For our particular example, ∂Input=|7.0 − 6.9|/7.0 = 0.0143. With "truculently" replaced by "belligerently" in a large corpus (100,000 documents in which "truculently" occurs 865 times in 617 documents), we measure the impact on accuracy for those 617 documents. Originally, 587 of the 617 documents were correctly classified; with the substitution of "belligerently for "truculently" in the documents, the accuracy dropped to 575 of the 617 documents (14 new ones were misclassified, and 2 of the formerly misclassified documents were now classified correctly, for whatever reason). Thus, ∂Accuracy = |587 − 575|/617 = **0.01945**.

The other part needed to calculate Resilience$_{\text{System}}$ (Equation (5.2)), as mentioned above, is ∂Input. Using the similarity of "belligerently" (0.9 to "truculently"), the error between these two terms is 1.0 − 0.9 = 0.1. Error needs to be normalized to the relevant sentence length. For these 617 documents, there are 11,679 words, of which 9143 are not stop words. Thus, we have a mean of 9143/617 = 14.82 terms to evaluate for similarity in each sentence. Eight hundred and sixty-five terms are changed with error = 0.1 each, so the total error is 865 × 0.1 = 86.5, and this is 86.5/617 = 0.140/sentence. The ∂Input = |14.82 − 14.68|/14.82 = 0.140/14.82 since the numerator is the error. This value comes to **0.00945**.

Plugging into Equation (5.2), Resilience$_{\text{System}}$ = ∂Input/∂Accuracy = (0.00945/0.01945) = 0.486. Based on this resilience, a change of input of 48.6% will result in 100% loss of accuracy. If there are N classes, the amount of input change to reach random guessing is given by Equation (5.5):

$$\partial \text{Input}_{\text{to reach random guessing}} = \text{Resilience}_{\text{System}} \left(\frac{N-1}{N}\right). \tag{5.5}$$

Here, N = the number of classes. The Resilience$_{\text{System}}$ value is therefore multiplied by $(N − 1)/N$ since that amount of change in input should reduce accuracy to random guessing: we assume nearly 100% accuracy to start with; otherwise, we can correct the $(N − 1)/N$ multiplier to account for this discrepancy. For our example, random guessing is 25% accuracy and thus

Table 5.3 Simple example for computing Resilience$_{\text{System}}$ by making a small change in the input. In this case, 617 documents with 865 instances of the word "truculently" have that word changed to "belligerently" and the effect on overall system accuracy measured. Similarity of the documents after substitution is 0.9905 compared to before substitution, and accuracy drops from 0.9514 to 0.9319. Using Equation (5.2), Resilience$_{\text{System}}$ = 0.486. There are many other variants for computing resilience possible, but the example given here is straightforward to compute and illustrates how resilience can be derived from modest changes in the input and measured impact on the output (in this case, accuracy).

Condition	Similarity to original document	Classification accuracy	Resilience
Original document with "truculently"	1.000	0.9514	Undefined
Documents with "truculently" replaced by "belligerently"	0.9905	0.9319	0.486

∂Input to reach random guessing is 0.486(0.75) = 0.3645. Once we have changed a little over a third of the words that are not stop words, we cannot discern the classes of the sentences any more. The results of this example are presented in Table 5.3. This assumes that the instantaneous rate of change of input with respect to accuracy does not change much over the range of change in input, which is, of course, not usually the case. The nature of that relationship entirely depends on how important the errors added are to the downstream accuracy (clustering, classification, categorization, search query matching, etc.) However, Equation (5.2) gives us a single metric to rate different systems. In general, we favor systems with higher values of Resilience$_{\text{System}}$. We especially favor them if we want them to work well in actual deployment.

So, we have shown that we can compute resilience relatively easily. We come back now to the intended breadth of the system. Breadth is a measure of resilience across the domain of the input. The input domain can be a gamut across any relevant variable on the input data; for example, location (spatial), time (temporal), operator (person collecting the data), organization, or even auxiliary measures like precision and recall of the data. We used temporal variability in our example (Figure 5.4) below. The resilience is the performance on the particular training set of interest. For the two algorithms compared, Algorithm B has the higher resilience since its value is 0.534 at the midpoint of the domain, while that of Algorithm A is 0.486 (Algorithm A is the one used in Table 5.3). However, when the same systems are compared against data they were not trained on, their resilience changed significantly. In general, the farther the new data is from the training set (x moves toward 0.0 or 1.0 on the domain axis), the lower the resilience. **Breadth** is defined as the

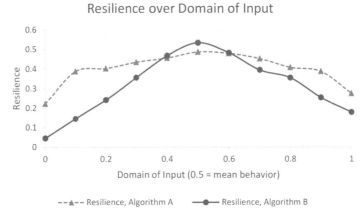

Figure 5.4 Plot of resilience (*y*-axis) against domain of the input (*x*-axis). Elements are assigned to the domain based on some factor that partitions the input. In the example given here, the training set is at domain $x = 0.5$, and $x = \{0.0, 0.1, 0.2, 0.3. 0.4\}$ are for $\{5,4,3,2,1\}$ time intervals earlier and $x = \{0.6, 0.7, 0.8, 0.9, 1.0\}$ are for $\{1,2,3,4,5\}$ time intervals later. Algorithm A has the greater Breadth (area under curve = 0.4136) compared to Algorithm B (area under the curve = 0.3337), but Algorithm B has higher resilience according to Equation (5.2) at 0.534 compared to 0.486 for Algorithm A.

consistency of resilience to new data, and for Figure 5.4, this is defined as the area under the curve. This area is 0.4136 for Algorithm A, and is only 0.3337 for Algorithm B. Thus, Algorithm A has greater breadth, but Algorithm B is more resilient.

Breadth and resilience are obviously closely related. We can think of breadth as system robustness; that is, a measure of how well the system will react to "unexpected" input. Resilience as defined in this section is more closely akin to "error tolerance."

5.5 Applications/Examples

Among the most important applications of text analytics are search and its flip side, information retrieval. For a large number of businesses, a familiar lament is often "Wow, if we only knew what we knew!" This should not be the case. Storage of information should be built with retrieval, not just archiving, in mind. In an *ideal information system*, the set of search queries used by someone hoping to find information in a text corpus should be an exact match to the set of keywords used to tag the documents. This provides us with a functional means of generating either or both of these sets of terms:

our optimized output is the one that maximizes the overlap between the search term and keyword sets.

5.5.1 Document Clustering

Document clustering, covered in Chapter 3, is also an approach that can benefit significantly from optimization approaches. The purpose of document clustering is to enable related information to aggregate so as to provide a more distinct set of descriptive terms than any of the individual elements can by itself. This is the Central Limit Theorem in action − a larger pool of related items will generally provide a closer estimate of the mean term distributions than a smaller pool. The standard error of the mean of a population is $\sigma/\sqrt{n}$, where σ is the standard deviation and n is the number of samples. Thus, as n increases, the estimate of the mean has a smaller range and, thus, a better estimate. One functional optimization for document clustering of note is for the cluster optimization algorithm to be guided such that the cluster descriptors (categories) best match the top search queries for the overall document set. This can be done quite simply by computing the Jaccard similarity index between the set of categorizing terms generated from each candidate set of clusters and the search queries, and selecting the cluster representation that maximizes this ratio of the intersection of the two sets over the union of the two sets.

5.5.2 Document Classification

Similarly, document classification can benefit from functional optimization. The class names are typically pre-defined, for example, from a taxonomy or other domain expertise-driven set of labels. However, they can also be automatically generated from the data sets assigned to the classes by domain experts. Functional optimization of classification, therefore, may be a little more subtle than for clustering. Here, the class memberships may be optimized not just so that the text statistics of the classes best match the name of the classes but so that the final set of classes provide the least additional categories over those provided by the class names. In other words, a principle of parsimony is applied such that class assignments result in the minimum number of new categories over those implied by the class names. This means that the final classification is a linear combination of the classification results and the "parsimony" algorithm (which can generally be performed by testing the assignment of the lowest certainty classified elements). This is described in Equation (5.6), where α is the regularization variable, and varies from 0.0 to 1.0. Generally, setting α in the range of $0.1-0.2$ suffices, depending on what percent of elements are evaluated with the parsimony algorithm (also

0.1−0.2 works as a good range there).

$$\text{Class} = (1 - \alpha)\text{Classification\_Result} + (\alpha)\text{Parsimony}. \qquad (5.6)$$

5.5.3 Web Mining

Another application space of note for text analytics is that of web mining; that is, of drawing meaningful inferences from data and text that is mined from online information. Web information typically has a different relationship to text analytics than does more structured information such as articles, books, and training material. However, even here, functional optimization approaches have value. One of the key aspects of online information are links; that is, connections between different elements online. The Internet provides a "fractal" design characteristic such that at many different scales, there is "small world" behavior where links are very strong amongst a smallish set of elements and relatively weak with the other elements in the same partition of elements. Small worlds are very important in text analytics and, in particular, for learning. Small "small worlds" can be used for intense training on a specific task; medium "small worlds" can be used for course curriculum planning, and large "small worlds" can be used for developing discipline mastery. In our small world example, we will consider all three sized "small worlds" together in an optimization problem.

One functional optimization approach for web analytics is to create small worlds at each of several scales such that they are actually fractals of small world behavior at the levels above and below them. For example, suppose that we have three scales, aggregating 100 documents in each partition at the lowest level, 1000 documents in each partition at the middle level, and aggregating 10,000 documents in each partition at the highest level. One simple way of assessing the "small world-ness" of each partition is to find what we term the "Pareto points" of linked behavior at each level. The usual presentation of the concept of a Pareto effect is often oversimplified to that of "80% of a task is completed by 20% of a team", and so if there is a Pareto effect for linking, it is that a vast majority of the links are to a much smaller percentage of possible linkage points. These **Pareto points** can be, for example, 10%, 20%, and 40% of the total links from one set of documents to all the other sets of documents, and what we are concerned with is the minimum percentage of other documents that need to be connected to achieve these levels of connection. The smaller the minimum set of documents connected to by 10%, 20%, and 40% of the links, the more tightly connected these "most connected" documents are. An example of three different document sets and their Pareto points are shown in Table 5.4. In this example, there are 10^6 (one million) documents in each of the Sets

Table 5.4 Table of *Pareto points* for three different networks at 10%, 20%, and 40% of their links. Links are assessed in partitions of 100, 1000, and 10,000 out of 1,000,000 documents (that is, there are 10,000, 1000, and 100 documents, respectively). Based on their lower variance, Sets B and C seem to behave more similarly at each scale than Set A. Please see text for details.

Partition (of 10^6 documents) and set	10% link mark	20% link mark	40% link mark
100 document partitions, Set A	0.0074	0.0253	0.0774
1000 document partitions, Set A	0.0041	0.0192	0.0633
10,000 document partitions, Set A	0.0113	0.0333	0.1197
Mean (Std.), Set A	**0.0076 (0.0036)**	**0.0259 (0.0071)**	**0.0868 (0.0294)**
100 document partitions, Set B	0.0112	0.0378	0.0874
1000 document partitions, Set B	0.0103	0.0314	0.1014
10,000 document partitions, Set B	0.0133	0.0417	0.1204
Mean (Std.), Set B	**0.0116 (0.0015)**	**0.0370 (0.0052)**	**0.1031 (0.0166)**
100 document partitions, Set C	0.0099	0.0274	0.1015
1000 document partitions, Set C	0.0116	0.0331	0.1194
10,000 document partitions, Set C	0.0145	0.0378	0.1342
Mean (Std.), Set C	**0.0120 (0.0023)**	**0.0328 (0.0052)**	**0.1184 (0.0164)**

A, B, and C, and so if we create 100 partitions, each one holds 10,000 documents; if 1000 partitions, each holds 1000 documents; and if 10,000 partitions, each holds 100 documents. Links between the documents were determined by matching the highest TF*IDF scores in all documents.

In Table 5.4, we see that 10% of the links form between a given partition and its most-highly connected 0.76%, 1.16%, and 1.20% other partitions. This level of connectivity is 13.2 (from 10/0.76), 8.6 (from 10/1.16), and 8.3 (from 10/1.20) times the mean level of connectivity, showing that the small world assumption (the basis of the Pareto) holds. All of these values are well above the "traditional" Pareto of 80/20, which assigns 4.0 times the expected value to the "small world." Also in Table 5.4, 20% of the links connect to just 2.59%, 3.70%, and 3.28% of the other partitions. This level of connectivity is 7.7 (from 20/2.59), 5.4 (from 20/3.70), and 6.1 (from 20/3.28) times the mean level of connectivity. Similarly, for 40% of the links, the 40% mark is achieved at 8.68%, 10.31%, and 11.84% of the other documents. From these results, we see that Set A behaves with more of a Pareto effect (it reaches 40% of its links with less overall connectivity; that is, to only 8.68% of the other partitions), but it has poorer scaling since it has the highest variance (represented by the Std., or standard deviation, in Table 5.4). In particular, it has the highest coefficient of variation, or ratio of Std./Mean. This value is 0.47 (10% link), 0.27 (20% link), and 0.34 (40% link) for Set A. The respective values are {0.13, 0.14, 0.16} for Set B, and {0.19, 0.16, 0.14} for Set C.

Which of these sets is optimal for small world aggregation? Not surprising given what you have seen so far in this chapter, it is conditional. Set A creates a stronger Pareto effect, with a stronger set of connections between related partitions. If you want smaller worlds, with presumably more reinforcement of topics, this is your set. If you, however, want the connectivity to scale across the size of the partitions, in order to have similar content relationships among the partitions, then Sets B and C are probably closer to your optimum.

5.5.4 Information and Content Extraction

Information extraction, or the ability to collect relevant facts and inferences about the relationships between them, usually is employed to upgrade the value of unstructured text. In many systems, this is how structure is added to text. The related task of concept extraction is focused on the grouping of text expressions based on their semantic similarity. Combined, these two approaches can be traded off each other for optimization in a two-step iterative process along the lines of E-M or competition–cooperation algorithms described earlier. The system is, for example, first mined to collect facts and inferences, and then those that do not match any of themes of the semantics are removed. Then, the semantic themes are generated and additional facts or information matching those themes are newly extracted. This two-step process occurs until stability is attained:

(1) Extract facts and inferences using statistical text analytics. Prune terms with poor match to semantic analysis.
(2) Use the output of the semantic analysis to find additional facts and inferences. At the end, prune the lowest weighted (lowest confidence) terms gathered by this method or by the method of (1).

5.5.5 Natural Language Processing

NLP is the primary set of approaches for generating the facts and inferences about the text. As discussed in Chapter 1, NLP includes part-of-speech tagging, word frequency analysis, and other computational linguistics. Optimization of NLP output can be functional when it is tied to the best overall performance of the output for specific text analytics tasks. These include search query behavior, topic association, indexing, and tagging (keyword generation) just to name a few. An interesting application of NLP toward functional optimization that has not, to our knowledge, yet been tested is to use NLP history information (text analytics over time) to automatically generate a **creation order** for a large set of documents. This is topically quite distinct from the determination of a reading order for learning, training, and other knowledge-collecting tasks.

5.5.6 Sentiment Analysis

Sentiment analysis is a potentially beneficial tool for augmenting traditional NLP and ML-based text analytics. One means of using sentiment analysis is as first decider for a downstream intelligent system. That is, if the sentiment is "happy," we may have found that using Algorithms A and B in parallel provides the highest system accuracy; if the sentiment is "angry," we may have found that using Algorithms C and D in series provides the highest system accuracy. Such an approach is an implementation of the predictive selection design pattern for intelligent systems [Sims13].

In addition to such "adjacency" roles, sentiment analysis has a primary role in generating *semantic sentiment*. For example, when someone says, "what a morning!" it can mean elation, exasperation, and perhaps some other emotions. Being able to correctly interpret the sentiment may be very useful in tagging the text element for later processing. This is effectively metadata, as it is data not explicitly garnered from the literal content of the text. A functional optimization of the extraction semantic sentiment is when it maximally improves the correct usage of the text element in other NLP, linguistic, and analytics workflows.

5.5.7 Native vs. Non-Native Speakers

This is no insult to people who speak English as a second language, just as I would take it as no insult coming my way when I am "conversing" in Spanish, French, or German: non-native speakers of a language are not simply challenged by vocabulary, but often more strongly by the idiomatic expressions, the speed and facility at which syllables are combined, and the contextual words that are not spoken as much as acted out. Clearly, the presentation of text, whether in written or auditory form, is dependent on the abilities of the reader or listener. In order to optimize text for the level of fluency of the audience, having the text pre-mapped to all of its synonymic (and non-idiomatic) substitution phrases is a means of accommodating the settings across the gamut of language competency. **Personalized optimization** of the language, with one example being the following sentence, perhaps written by a young author trying too hard:

> *The iridescent albedo of the pavement gave forth a pale beacon into the lugubrious firmament.*

Now if the words and expression have been pre-mapped to simplifying (vocabulary condensing) synonyms, we can write the same sentence as

> *The scattered reflection of the street shone a pale light into the sad sky.*

In other words, the words are replaced with the synonym most likely to be understood by a non-native speaker. It may not be as poetic (well, maybe in this case, it is more poetic!), but it will be understood. The concept of personalized optimization is, of course, likely to become a more common approach as advanced artificial intelligence begins to add more realistic semantics to the speech and text generation processes.

5.5.8 Virtual Reality and Augmented Reality

One last application area should be mentioned here. Virtual reality (VR) and augmented reality (AR) are becoming more popular means of learning, training, and exploring with each passing year. This is another technology which will increasingly benefit from personalized optimization. Providing textual cues, for example, as signage, as subtitles, or as other linguistic content in AR/VR must consider a number of factors in order to be optimized for an individual. Among these are the following:

(1) Speed of advancement of words if presented as scrolling text − one of the key aspects of VR/AR success is not to have the text distract from the feeling of "presence."
(2) Position, font size, and color of subtitles to ensure readability against the background.
(3) Rate of change of other video content if, for example, the user suffers from cybersickness (nausea, disorientation, etc.) so as not to overwhelm the sense when text is provided.

Text is an important part of learning. Optimizing its integration into AR/VR is likely to be a key "human factors" research area long into the 2020s.

Table 5.5 COV comparison for three processes at times $t = 1$, $t = 2$, and $t = 3$. While the mean of the three processes does not change much over these times, there is a sudden decrease in the COV of process A and a corresponding increase in the COV, which together may be indicative of instability in the system at time $t = 3$.

Time	COV of process A	COV of process B	COV of process C
$t = 1$	0.82	0.84	0.83
$t = 2$ (consistent)	0.81	0.85	0.84
$t = 3$ (inconsistent)	0.76	0.92	0.85

5.6 Test and Configuration

Test and configuration is an important part of optimization, and, in particular, for functional optimization approaches such as personalized optimization and when comparing and contrasting multiple optimization strategies. Test and measurement of different optimization possibilities is, of course, no different than any other classification, information retrieval, or pattern recognition problem: we are concerned with the accuracy, the precision, the recall, the F1 score, the drift of performance, and the error measured during the training and validation stages. In many cases, the COV (coefficient of variation, or ratio of standard deviation to the mean) is a key measurement to make for testing consistency of an approach. This was exemplified by the three sets of documents evaluated for small world behavior in the previous section.

In addition, COV consistency can be used to show when input has likely changed in behavior such that the training data might need to be updated. This is shown in Table 5.5, in which the sudden drift in COV at time $t = 3$ indicates a change in processes A and B but stability in process C. A benefit of COV, like resilience in Equation (5.2), can provide a single metric to track for sudden changes in system input behavior.

Functional testing offers a different approach to optimization than traditional mathematical models, statistical inferences, and large system simulations. With functional approaches, often the decision on which of the candidate optima to deploy is in the adoption rate by users, or feedback from simple A/B or Likert tests. These types of feedback engage human factors experts, user interface designers, and professional testers, which might be folks unexpected to contribute to the optimization of the design of a machine intelligence system. However, the purpose of functional text analytics is to open up a system to all of its involved personnel. Customers and software testers alike can report bugs, complain about missing features, and otherwise relatively rate different system designs. Inference about the relative merit of each optimization candidate can be made by simply monitoring the traffic rate of good and bad recommendations, comments, and feedback. Ultimately, optimization is a matter of opinion.

5.7 Summary

This chapter describes how optimization is conditional. There is no single means of optimizing even simple text analytics problems. In comparing the impact of compression on document classification by two classification engines, we showed that there are at least four reasonable optimization strategies. In our example, the perhaps most surprising strategy (using a combination of the validation output for both classification engines) provided the highest accuracy on test data. We then showed how defining an "optimal" optimum is possible, in this case, using the product of the optimum's deviation from mean behavior multiplied by the size of the neighborhood converging on the optimum. Iterative means of optimization, including E-M and competition–cooperation, were then described. A means of automatically terminating an "optimal clustering" approach was defined. In ML, concerns about the quantity, quality, breadth, and depth of training data were overviewed. Next, we addressed system and design considerations for optimization. We described the geometric increase in the level of complexity for optimization when a pipeline of processing steps is involved. We next provided a quantitative means of defining system resilience based on creating a small error in the input and measuring the resulting error in a system metric such as accuracy. The means of comparing system resilience to system breadth was then defined. Functional means of assessing clustering, classification, and web mining emphasized the many choices for optimization. With web mining, the ability to measure connectedness ("small world" behavior) offered several choices to the application designer. Other applications of functional optimization were overviewed briefly. The chapter concludes with a brief discussion of testing for optimization.

Was this chapter written in an optimal way for your needs? Certainly, you will appreciate by this point of the chapter that optimality is conditional. If you are willing to accept the premise that optimality is, in some respects, an illusion, then you are probably in at least partial agreement with the recommendations of this chapter. If not, well, there may well be a second edition to this book someday, and you can hope for a different optimization therein.

References

[Abby20] Abbyy Technology Portal, Supported OCR Languages, accessed 18 July 2020, at https://abbyy.technology/en:products:fre:win: v11:languages.

[Demp77] Dempster A, Laird N, Rubin D, "Maximum likelihood from incomplete data via the EM algorithm," Journal of the Royal

Statistical Society, Series B (Methodological), vol. 39, no. 1, pp. 1-38, 1977.

[Frey07]   Frey BJ, Dueck D, "Clustering by passing messages between data points," Science 315 (5814), pp. 972–976, 2007.

[Geye92]   Geyer CJ, "Practical Markov chain Monte Carlo," Statistical Science 7(4), pp. 473-483, 1992.

[Gret19]   Gretter R, Allgaier K, Tchistiakova S, Falavigna D, "Automatic Assessment of Spoken Language Proficiency of Non-Native Children," in ICASSP 2019, DOI 10.1109/ICASSP.2019.8683268, 2019.

[Like32]   Likert R, "A Technique for the Measurement of Attitudes," Archives of Psychology 140, pp. 1–55, 1932.

[Luhn59]   Luhn HP, "Keyword-in-Context Index for Technical Literature (KWIC Index)," American Documentation, Yorktown Heights, NY: International Business Machines Corp. 11 (4), pp. 288–295, 1959.

[Sims13]   Simske S, "Meta-Algorithmics: Patterns for Robust, Low-Cost, High-Quality Systems", Singapore, IEEE Press and Wiley, 2013.

[Yaco03]   Yacoub S, Simske S, Lin X, Burns J, "Recognition of Emotions in Interactive Voice Response Systems," in EUROSPEECH-2003, pp. 729-732, 2003.

6

Learning

"Development is a series of rebirths"
– Maria Montessori

"Expecting all children the same age to learn from the same materials is like expecting all children the same age to wear the same size clothing"
– Madeline Cheek Hunter

"Learning is not attained by chance, it must be sought for with ardor and attended to with diligence"
– Abigail Adams

Abstract

Learning is the process of ingesting information with the purpose of retention. Learning, and its associated topics of teaching, studying, rehearsing, and cramming, is a thoroughly functional element of text analytics. In many ways, it is the culmination of functional text analytics, where machine learning pays off as a means of engendering human learning. In this chapter, the perspective of learning as the sequential flow of content will be advanced. The implications of this for machine learning (for example, bias toward redundancy) and for test and measurement are described. Applications of sequential flow of text analytics including reading order optimization, curriculum development, customized education planning, and personalized rehearsing.

6.1 Introduction

The beginning of text analytics, according to many philologists, was for rather mundane purposes. Early recorded history is inescapably poorly understood because, of course, it extends back only so far as the medium for the text (clay tablet, papyrus, cave wall, animal bone, etc.). But, from what we can glean from the artifacts that have survived, early writing was used for recording transactions. In other words, the first writing was a database, and clay and papyrus were tablet-based devices to access the database. In other words, they were transaction servers. Yes, we are pushing the analogy a bit here, but keep in mind that the Ancients were far more sophisticated than some anthropologists had earlier thought. After all, these same databases in Ancient Egypt included 18 relational databases (that is, papyri) that relate "prescriptions for disorders of the teeth and oral cavity, seven are for remedies to prevent tooth loss by packing various materials in paste form around the tooth and the surrounding gums" [Fors09].

Clearly, text has been employed since its early days as a means to provide archiving, authority, and some level of authentication. Text and its glyphs have transparency: anyone having literacy can interpret it, and society as a whole agrees on its encoding and decoding process (syntax) along with its meaning (semantics). Clay tablets, like dinner plates, incurred some expense to create, impress, and fire, meaning they provided an authoritative record of the transactions. Physical databases such as papyrus and clay tablets, just like paper and tapes, are important in establishing veracity. Both personal (e.g., signature) and natural (e.g., the shape of the impression made in clay) forensics have historically been used for authentication. When data is saved and preserved (that is, archived), trusted and validated (that is, authoritative), and is proof of information in the physical world (that is, authenticated), it provides a certification of information. A certification, even today, is *proof of learning*.

Thus, text is used for all aspects of learning. Learning is the ultimate text analytic: the text has been abstracted into neural "data" within the brain of the learner, and the learner's ability to recall the information is the *functional measure* of the text analytic. Taking a test and passing, from this perspective, is a form of functional text analytics. When important information that is intended to be shared is written (that is, presented at least in part as text), this is training text. Someone who is forced to learn from this is engaged in "cramming," which is the process by which as much text as possible is internalized (into at least medium-term memory) in order for the functional analytic of conveying the correct answers on an examination to achieve the requisite accuracy. The term here makes sense: "cramming" implies overfitting, which might mean no long-term changes in the vast cerebral cortex is accomplished. More unhurried internalization of the text is "studying," with the intent being that these memories are more permanent; that is, associated with long-term memory. The fact that studying and student have the same root also makes sense: a student studies. Focused studying, wherein specific points are preferentially learned, has another name. Converting short- and medium-term memories into long-term memories is typically done through the process of rehearsing. Rehearsing, in an analogy to classification, is a form of boosting, wherein a given set of data is emphasized to train the system (here, the human brain) to be more accurate. In another analogy to text analysis, we can think of rehearsal as a means of increasing the weighting of given terms; effectively, increasing their term frequency.

In many ways, the object of educators − those who teach, train, and certify people − is to produce an archival representation of a corpus that encourages its reuse. Why do we add tags, indices, cluster assignments, and class assignments to text elements? It is because all of them provide additional access points for the corpora. Search queries work far better on text elements that have already had meaningful and accurate metadata added to them. Adding relevant functionality, or utility, to an already-existing database makes a large corpus more suitable for training and learning purposes.

Ever since 1998, automatic paper-recommender and book-recommender [Alha18] systems have been a focus of study, especially for learning purposes. In one highly cited recommender system survey article assessing more than 200 research publications, 55% of the systems used content-based filtering, or CBF (wherein users are interested in items that are similar to those that were previously liked). TF*IDF (term-frequency times inverse document frequency) was the most commonly used weighting scheme [Beel16]. Collaborative filtering (CF) is another popular method for recommendation systems, which is similar to CBF except that recommendations are based on the preferences of other users [Taru18]. For example, Netflix will recommend a series of videos to watch next based on what other users who watched the same show also watched. This is purely correlation, and any actual personalization of this approach is fortuitous but not causal. One of the biggest issues with both the CBF and CF approaches is that, frequently, there is not enough information about the users and their preferences, which can lead to a problem of over-specialization. Additionally, new areas of growth are never suggested [Khus16]. CBF and CF may be good routes for finding a narrow set of documents for a learner to consider, but they do not take into consideration the appropriate order for those documents to be presented for the most effective learning outcomes.

6.1.1 Reading Order

Learning in the large requires the proper sequencing of many documents in order to enhance understanding and retention by the reader. In general, for learning (and especially the fast-growing remote learning field), we are interested in creating a list of documents to provide a reader in a specific sequence, with the "next best document" to be provided once she has finished with the current document. In general, the job of the text analytics expert is to determine a document sequence in order to ensure that the ND provides the right amount of overlap for reinforcement, and simultaneously, the right amount of non-overlap to encourage the ingestion of new concepts and facts.

Accurate sequencing of information becomes even more difficult when summaries of content, rather than the initial content itself, is being used. This is more difficult because the consequences of peeling away much of the nuance in the summaries include poor retention, rough transition between documents, and loss of the author's intended context. Summarization, however, is essential for many tasks since time and other forms of input (slides, video, augmented reality (AR), etc.) have become increasingly important. Summarization can be thought of as a spatial representation of a corpus. Each document that is summarized occupies a spot in the overall mapping of the corpus content. If the content is evenly represented in the summary, then the summary is effectively a shallow, spatial representation of a document. Learning, however, is concerned with the temporal representation of the corpus. Thus, the manner in which learning is to occur (temporally) is structurally incompatible with how summarization occurs (spatial). What does this mean for summarization and learning? You likely saw this coming – summarization can be *functionally optimized* to support learning by comparing the reading order obtained when applying a reading order operator to the original set of documents to the reading order obtained on a summarized set of documents.

There are multiple methods of generating reading orders, and some of the important factors in their determination are discussed below. However, the most important task in determining a reading order may well be the starting document. This document often can be determined as the document that is most central to the overall content. One simple means of assigning the most central document is to determine the one that has the highest overall cosine similarity to all of the other documents in the corpus [Oliv18]. A simplification of the normal cosine similarity is possible, where the similarity is emphasized by the dot product of each term frequency (tf; the percentage of all words) value for each word in the overall corpus (set of all documents). The Starting Document Index is the index for which the sum of the term frequencies multiplied together is highest, as per Equation (6.1).

$$\text{Starting Document Index} = \arg\max_{i=1...N_{\text{documents}}} \sum_{j=1,\, j\neq i}^{N_{\text{documents}}}$$
$$\times \sum_{w=1}^{N_{\text{words}}} \text{tf}(w, i) * \text{tf}(w, j). \qquad (6.1)$$

Table 6.1 Reading order example. The reading order of the original document set is determined, and the reading sequence is recommended as documents $\{1, 2, 3, 4, \ldots, 10\}$ in order. The same reading order algorithm is used on two different summaries of the documents, and the reading orders aligned with the original. The weighted difference is calculated from Equation (6.2). Here, it is clear that the second summary performs more like the original documents for reading order.

Reading order, original text	Reading order, summary (1)	Weighted difference (1)	Reading order, summary (2)	Weighted difference (2)
1	2	$10 \times \|2-1\| = 20$	1	$10 \times \|1-1\| = 0$
2	4	$9 \times \|4-2\| = 18$	3	$9 \times \|3-2\| = 9$
3	1	$8 \times \|1-3\| = 16$	2	$8 \times \|2-3\| = 8$
4	5	$7 \times \|5-4\| = 7$	4	$7 \times \|4-4\| = 0$
5	3	$6 \times \|3-5\| = 12$	7	$6 \times \|7-5\| = 12$
6	8	$5 \times \|8-6\| = 10$	6	$5 \times \|6-6\| = 0$
7	6	$4 \times \|6-7\| = 4$	5	$4 \times \|5-7\| = 8$
8	7	$3 \times \|7-8\| = 3$	9	$3 \times \|9-8\| = 3$
9	11	$2 \times \|11-9\| = 4$	8	$2 \times \|8-9\| = 2$
10	9	$1 \times \|9-10\| = 1$	10	$1 \times \|10-10\| = 0$
Sum of differences (Equation (6.2))	N/A	95	N/A	42

In Equation (6.1), the important variable is N_{words}. Usually, this is well less than the full complement of words in all of the documents. The stop words (the, is, and, or, etc.) are usually dropped from the list since they do not contribute to the meaning of the corpus. In addition, only words which occur with higher frequency than T times their overall frequency in the language of the documents may be included. Regardless, once the most central document is found, we can sequence the rest of the documents through a number of different methods, as will be discussed shortly. Before getting to that, however, let us explore some of the *functional motivation* for being able to derive reading orders of text corpora by showing how to compare the reading orders obtained from an original and summarized sets of documents (Table 6.1).

Table 6.1 provides the reading order output for an original set of documents and compares it to that of two different summarizers. Once the reading orders are obtained, the weighted difference between them can be computed using Equation (6.2).

$$\text{Weighted Difference} = \sum_{n=1}^{N_{\text{ranks}}} (N_{\text{ranks}} - n + 1) \left| \text{Order}_{S,n} - n \right|. \quad (6.2)$$

Here, N_{ranks} is the number of relevant reading order ranks. In Table 6.1, this number is 10. Since the learning is more heavily dependent on what is read first, here we use a weight of 10 for the first document that is (N_{ranks}

$- n + 1) = 10 - 1 + 1 = 10$ when $n = 1$. For the 10th and final ordered document, this weight is $(N_{\text{ranks}} - n + 1) = 10 - 10 + 1 = 1$. This weight is multiplied by the absolute value of the difference in the reading orders for the summarizer compared to the original; that is, $|\text{Order}_{S,n} - n|$, where $\text{Order}_{S,n}$ is the reading order placement of document n in the summary (S). Thus, if $n = 7$ and the document that was in reading order "5" for the original documents is the recommended 7th summary to read, $\text{Order}_{S,7} = 5$. In Table 6.1, the original reading order of the full (non-summarized) documents is $\{1, 2, 3, 4, 5, 6, 7, 8, 9, 10, 11, 12, \ldots\}$. The documents are thus labeled with the suggested reading order of the reading order engine (ROE). Two different summarization engines (see Chapter 3) are then used to summarize all of the documents, and the ROE is used to suggest the reading order of the summaries. The first summarization engine's output (second column, Table 6.1) is analyzed by the ROE and the suggested reading order is the set of 10 documents $\{2, 4, 1, 5, 3, 8, 6, 7, 11, 9\}$ in that order, which results in a sum of weighted differences (third column, Table 6.1) of 95. A second summarization engine's output (fourth column, Table 6.1) is analyzed by the ROE and the suggested reading order there is the set of 10 documents $\{1, 3, 2, 4, 7, 6, 5, 9, 8, 10\}$ in that order, which results in a sum of weighted differences (fifth column, Table 6.1) of 42, less than half of summarization engine 1. In this case, the sum of absolute differences, Equation (6.3), is also much higher for summarization engine 1 (at 15) than for summarization engine 2 (at 8). There are other ways to weight the differences, including placing higher multiples at both the start and the finish to ensure that the material "ends strongly." However, Equations (6.2) and (6.3) represent two extremes for determining a sum of differences.

$$\text{Sum of Absolute Differences} = \sum_{n=1}^{N_{\text{ranks}}} |\text{Order}_{S,n} - n|. \tag{6.3}$$

Having established that we can compute the weighted difference or sum of absolute differences between two reading orders, we now return to the manner in which the reading order is fully computed. This boils down to the question of what to do when a reader completes her reading of a document. Which is the next one to read? If the first document is determined as the one most central to the content, then the next one is selected to have some overlap with the first one, and some new topics. How much of these two? The answer, not surprisingly, depends on a number of factors, which include but are not limited to the following factors, each of which can be considered separately as described here, or together, as will be described after we have considered all five separately.

(1) *Proficiency of the reader:* In general, the higher the proficiency of the reader, the more aggressive the reading order algorithm can be. This

means the amount of overlap with all of the previous documents of the ND to be selected is inversely proportional to the proficiency of the reader. Proficiency itself can be updated after each document is completed with, for example, a test of proficiency. Historical data can also be used to weigh in here. If, for example, a neophyte is shown to do best with a document that has 85% topical overlap with what has already been read, and a true proficient does best with 55% overlap, then someone of middling ability might do best with 70% overlap.

(2) *Number of documents in the same cluster:* The set of documents that are candidates for the ND to be read can be clustered based on their relative distances. The "inter-document distance" between any two documents can be determined by Equation (6.4):

$$\text{Inter} - \text{document Distance}\,(i, j) = \sum_{n=1}^{N_{\text{words}}} \left(tf\,(n, i) - tf\,(n, j)\right)^2.$$

(6.4)

Once these distances are computed, the documents can be linked through agglomerative hierarchical clustering (AHC) [Day84], where the threshold for intra-cluster distance is allowed to rise over time. All the documents in a cluster surrounding the currently-being-read document (CBRD) are then assessed for similarity to the current set of documents that have been read, and the document in the same cluster as the CBRD (lowest inter-cluster distance) that is the farthest away from the current set of documents that have been read (highest intra-cluster distance) is the ND selected in the sequence. This has a familiar tradeoff to it, along the lines of the expectation−maximization and competition−cooperation approaches of previous chapters. Selecting the same cluster as the CBRD keeps the ND from straying too far, while selecting the farthest away document from the CBRD keeps it from staying too close. In the next iteration, only the new currently-being-read and unread documents are formed into clusters. In this way, new clusters can potentially form, allowing the reading order to break out of the current cluster without having to completely exhaust it. In short, the steps are as follows:

(a) Form document clusters out of the CBRD and all documents not yet assigned to the reading order. Any documents already assigned to the reading order sequence in previous steps are no longer part of the clusters. Using AHC or regularized k-means clustering (RKMC) [Sims19] allows the "cut point" on the tree for AHC, or the maximum intra-cluster distance for RKMC, to slowly increase as the number of remaining document increases. In other words, the product of this distance and the number of remaining documents is a constant.

(b) Select the document in the same cluster as the **CBRD** that is the maximum distance away from the CBRD as the **ND** to be read.

(c) Make the ND the CBRD, and return to Step (a) until the desired number of documents are in the sequence; if no documents remain; or if no documents remain that are a suitable distance away from any that have already been read.

(3) *Number of documents in the same class:* For this approach, the documents are classified using an existing document classification approach (e.g., any of the methods described in Chapter 3). The steps involved are:

(a) Within a document class, assign the document farthest away from the CBRD as the ND.

(b) Update the ND to be the CBRD.

(c) Test if the closest document to the new CBRD is still within the same class. If it is, then return to (a) if condition (e) below is not met.

(d) If the nearest document to the new CBRD is in another class, then pick that closest document as the ND, and reset the current class and the CBRD to be the ND and its class. Then return to (a) if condition (e) below is not met.

(e) Before returning to (a), check if the desired number of documents are in the sequence; if no documents remain; or if no documents remain that are a suitable distance away from any that have already been read. If any of these conditions are observed, the reading order sequence is completed.

(4) *Number of documents with the same categories:* For this approach, instead of term frequencies, features, or classes, the documents already have tagging, indexing, or other metadata content that can be used to determine document similarity. Here, the similarity is simply a weighted sum of the matching terms between each document, as per Equation (6.5):

$$\text{Document Similarity}(i,j) = \sum_{n=1}^{N_{\text{metadata}}} W(\text{metadata}(n))$$
$$\times (\text{Exists}(i) * \text{Exists}(j)) \qquad (6.5)$$

Here, $W(\text{metadata}(n))$ is the relative weight assigned metadata(n); $\text{Exists}(i) = 1$ if the metadata(n) exists in the metadata of document(i), and =0 if it does not; and Exists(j) = 1 if the metadata(n) exists in the metadata of document(j), and =0 if it does not. After that, reading order is determined as for documents in the same cluster, or (4) above.

(5) *Number of documents the reader has time to read:* This constraint is a bit different than the first four. Here, the limiting factor may well be the time available for the reading. This can set a limit on the number of documents, the total page length of all of the documents, the number of words in all of the documents, or some combination of two or more of these constraints. Therefore, this constraint is largely upstream of the other four, and effectively tells us how many documents we should be concerned with sequencing. For example, the choice of 10 documents to sequence in Table 6.1 came from a consideration of the number of documents that can be read in the anticipated time period the reader has to ingest the material.

A robust reading order algorithm can consider any five of these factors separately, as shown here, or together. For example, the proficiency of the reader can be factored in. A neophyte might read the 10 documents in Table 6.1, while someone with mid-level proficiency might be able to read 12 documents if they are rated as having the same difficulty. A skilled professional might be able to read 15 documents of the same difficulty. However, if the skilled professional is assigned documents as challenging to her as the 10 documents are to the novice, then the highly proficient person still is assigned 10 documents. Needless to say, there is room for new findings in this area of research.

6.1.2 Repurposing of Text

Another area of current research interest is text repurposing. Learning represents a vast opportunity for additional research in text repurposing. Modern documents undergo a high rate of change themselves (particularly web documents, but also extending to books and magazines which are almost universally available in streaming form). Content is shared across many networks, particularly in news and social media, and disaggregating the content can be used to provide a form of "micro-reading order" which can be sequenced as described in the previous section. However, repurposing text offers several interesting research opportunities.

The first research opportunity is in text archiving. When content is reused, it is very important to have its original appearance properly acknowledged and preserved. The Internet itself benefits from the "Wayback Machine," which is an archiving service for the entire Internet (except for China and Russia), and can be accessed at *web.archive.org*. But, when smaller amounts of text are stitched together to create content for learning, providing proper acknowledgement is more difficult. It is the responsibility of the education/training provider to properly cite their content, but when the content is repurposed to a different format (for example, bullets on a PowerPoint slide or text to speech), this is often overlooked. In a world where sampling and "cutting and pasting" are commonplace and where copyright violations are more difficult to pursue (e.g., due to globalization and the transnational creation and digestion of content), the original source of the material may be difficult to discern. This is unfortunate for reasons other than the copyright issue. How can the validity and the authority of the content be determined? Clearly, this is an area of interest for text archiving professionals but is beyond the scope of this book. However, from a functional text analytics perspective, what may be important here is showing how the presentation of different content for training has varied over time. This includes potential reversal of the "metadata" associated with the content, which includes interpretation of its veracity and sentiment around it. For a "fact" that has changed in veracity over time, think of the one that "Columbus discovered America" (obviously not, since the Native Americans and Norse were there earlier, and he never realized that it was a new continent). For sentiment, think of the facts about Stalin as related to the denizens of the Soviet Union before and after 1953.

The main purpose of repurposing (sorry for that!) for this chapter is to optimize the reuse of existing materials to help someone understand a topic or skill better. How can we measure this? Automated testing has some value here, but mastery of a subject generally requires more in-depth evaluation than multiple choice answers can provide. There are the more difficult, and no doubt more noble, ways of determining improvement in proficiency, including grading short answers and essays. These require expense in both human expertise and human time. We are looking, instead, for functional means of determining expertise, and it is here that the reading order approaches we described above can be of additional value. For example, if, as the documents are ingested, the speed at which they are read stays the same or increases, we have evidence for the reader to be ingesting the material. If new skills associated with the materials are displayed – for example, the reader is able to use new equipment or otherwise show evidence of newly obtained mastery – then we can consider the material much better learned than otherwise. There are many different ways to map learning from ignorance to proficiency, and our intent in a training program is to maximize the amount of proficiency one gets over time. To that end, one useful functional measure of proficiency may be a consistent reading rate as new material is provided.

6.1.3 Philosophies of Learning

The functional measurement of learning should be tied to the learning goals. Given the many different philosophies of learning, we will recount several here to illustrate the breadth of possibilities. The first method to discuss here is the Waldorf, or Steiner, education program. This is eponymous to Rudolf Steiner, who named the teaching philosophy associated with his learning program Anthroposophy, or "human wisdom." As aficionados of hybrid methods for machine learning, we are particularly enamored with the integration of artistic, intellectual, and practical skills that are foundational to this method. Anthroposophy rightly asserts that integrating many different forms of learning simultaneously enhances creativity and imagination. As such, a functional measurement of success in this philosophy of learning is the creative output of the student. Since Steiner educators have a large amount of independence (consistent with the emphasis on creativity!) in their teaching curriculum, they are generally successful in creating functional measurements of student progress that are also tightly integrated into the holistic activities of the students. As one might imagine from this, standardized testing is not a huge part of the Steiner education (it is performed, if at all, in order to aid their students in their post-Steiner world). We see from this that the Steiner educational philosophy is logically consistent with a functional approach to learning, in which reading proficiency (for example, of instructions on how to build a balsa glider) can be directly related to the length of time the balsa glider is able to stay in the air after launching.

A separate, but related, learning philosophy is the Montessori school, also an eponym for its founder, Maria Montessori. Here, initiative and natural abilities of the student are the focus of the curriculum. The method is focused on what is called "practical play." Tied directly to functional measurements of learning, the Montessori students develop at their own pace. The method is tied to specific developmental ranges; for example, $2-2.5$ years, $2.5-6$ years, and $6-12$ years. The younger two age groups focus on sensory input, directly handling materials and objects in their environment. The latter age group is introduced to abstract concepts and, thus, can be expected to connect reading input to specific advancement in other skills. We can see that the Montessori method has analogous functional learning attributes to the Steiner method.

The third method we are overviewing in this section is that of Madeline Cheek Hunter, one of the people quoted at the start of this chapter. Madeline's teaching philosophy is called the "Instructional Theory into Practice" teaching model. It is a very popular model for education and was used in thousands of schools in the United States. In this section, we focus on three lists of seven components that Madeline used to view education.

The first is her set of teaching components, which comprise (1) knowledge of human growth and development; (2) content; (3) classroom management; (4) materials; (5) planning; (6) human relations; and (7) instructional skills. This set is comprehensive in that it accounts for how the classroom can best relate to the humanity of the students, and it also accounts for the pragmatic aspects of managing multiple students at a time. These led Hunter to a direct instructional model that she also summarized in seven components: (1) objectives; (2) standards; (3) anticipatory set (attention getter, summary of prior learning, preparation for the day's learning); (4) teaching (including checks for understanding and addressing feedback); (5) guided practice/monitoring (application of the lesson); (6) closure ("this is what we learned"); and (7) independent practice (homework). In this philosophy, the functional measurement of the teaching can be directly tied to the student's performance on the "check" for the lecture, the guided practice, and/or the homework. This affords ready personalization and, where properly designed, functional measurement of the learning. Hunter also incorporated key elements of these two sets of seven components into her instructional model, which incorporates both teaching and behavioral components: (1) objectives; (2) set, or hook, to garner student interest; (3) standards and expectations (the "rules"); (4) the teaching itself (inputs, demonstration, direction giving, and checking for understanding); (5) guided practice; (6) closure; and (7) independent practice. These are clearly either the same or analogous to the direct instructional model.

What these three different learning models have in common is that teaching is a closed-loop system. Even though there are familiar quotes about how a wise person learns from their mistakes, but the even wiser person learns from the mistakes of others, all three of these systems for learning acknowledge the fact that, in general, learning proceeds best when the student is consistently being expected to provide evidence of functionality with their learning. We could not agree more.

6.2 General Considerations

Maria Montessori notes that development is a series of rebirths. In French, rebirth is "renaissance," and a Renaissance person, or polymath, is someone who has an admirable skillset in a number of areas (usually bridging from the science and engineering side over to the liberal arts). In order to truly learn a new topic of breadth sufficient for learning, one has to be reborn to the material. This opens the door to a host of indirect means of assessing the successful incorporation of reading material. As one example, a direct comparison of the vocabulary of a learner with the material they are reading

Table 6.2   Comparison of the vocabulary proficiency of three learners at the start of training and after each successive third of their training. Learner 1 is assigned Novice status, Learner 2 is assigned Moderately Skilled status, and Learner 3 is assigned Expert status at the start of their learning. Please see text for details.

Compared vocabulary	Learner 1	Learner 2	Learner 3
Novice (at 0% training)	**0.64**	0.94	0.99
Moderately skilled (at 0% training)	0.42	**0.59**	0.83
Expert (at 0% training)	0.23	0.43	**0.54**
After 33% training	0.75 (+0.11)	0.61 (+0.02)	0.53 (−0.01)
After 67% training	0.79 (+0.15)	0.66 (+0.07)	0.57 (+0.03)
After 100% training	0.85 (+0.21)	0.72 (+0.13)	0.56 (+0.02)

(Table 6.2) may provide one potentially valuable means of assessing their progress. In Table 6.2, the "vocabulary proficiency" is determined from the comprehension in reading and the adoption in writing of the terms consistent with novice, moderate, and advanced (proficient) skill levels in a subject. This was assessed based on the vocabulary they incorporated (correctly) in a set of short answers writing assignments. In Table 6.2, Learner 1 was shown to have higher than 50% comprehension and adoption, or "proficiency," of the novice level (but less than 50% for moderate or expert proficiency) and, so, was assigned to the Novice class. Similarly, Learner 2 was assigned to the Moderately Skilled class, and Learner 3 was assigned to the Expert class.

The results of training differ for these three learners. Learner 1 very quickly improved their vocabulary proficiency in the Novice material assigned, moving from 64% to 75% proficiency with the Novice material after reading 1/3 of the training materials. While the rate of improvement slowed, the trend was continually upward, and, thus, training appeared to be working. In data not shown in Table 6.2, Learner 1 also improved vocabulary proficiency for Moderately Skilled (to 0.51) and Expert (to 0.32) proficiency vocabularies after training was completed. Thus, it seems reasonable to conclude that the training improved Learner 1's proficiency. Learner 2 was assessed as having higher proficiency at the onset of training and, so, was assigned to the Moderately Skilled proficiency level. Learner 2 showed less rapid progress, improving in vocabulary proficiency by a likely statistically insignificant +0.02 after the first third of training. However, in the latter 2/3 of training, Learner 2 improved another +0.11, even more than the +0.10 for Learner 1 in the same 2/3. Thus, Learner 2 may have struggled at first (admittedly we do not know the margin of error here), but then clearly showed progress with the learning, and by the end of the training had even improved vocabulary proficiency in the Expert category to over 50% (data not shown). The final learner highlighted, Learner 3, was assessed as an Expert (very high score for Moderately Skilled, and above 50% for Expert) but did not respond well to training throughout the experiment, ending (56%) with a vocabulary proficiency very similar to the starting point (54%). Overall, we can see several interpretations for the findings of Table 6.2, including the following:

(1) Since Learner 2 was slower to learn than Learner 1, and Learner 3 did not improve during training, it may well be that ~60% proficiency is the right score to be assessed as belonging at a particular level. Proficiency at the ~55% level may be an indicator of a mixed proficiency (e.g., Learner 3 should have been assessed as intermediate to Moderately Skilled and Expert).

(2) We do not know the margin of error, but based on the overall consistency of the story of the three learners, it is likely relatively small.

(3) The vocabulary proficiency score seems to be a good predictor for actual training proficiency. The correlation between assessed proficiency level and ability to improve proficiency with training is clearly high.

Vocabulary proficiency is only one of several proficiencies that might prove useful in assessment. Speed of test-taking, in addition to the test score, is also a likely indicator. Confidence scores, independent of performance scores, could also be a useful means of assessment, depending on the nature of the training subject(s).

Turning to our second quote, Madeline Cheek Hunter notes that children are just as varied in their learning aptitudes as they are in size and shape. Some learn from reading, some from writing, some from watching, and some from doing, and others from any combination thereof. Madeline thought that the major role of the teacher was to make decisions, and, in this context, the important decisions are those that lead to customization of the teaching curriculum for each student individually. In other words, the teacher is largely responsible for selecting the customized curriculum of the individual learner. With all the different media, content, learning philosophies, and perspectives that an educator has to choose from, the job of deciding has never been more difficult. This is, in fact, a perfect opportunity for text analytics to assist the educator. It is unlikely that we will ever want the task of education to fall entirely into the hands of an automaton (including algorithms and analytical approaches), but having them give an assist to the educator certainly makes sense. How often have we heard an educator, in exasperation or in knowing acceptance, lament, "I can tell nothing is getting through to you right now?" This comment is, properly, a launching point, not a sinking. Also, this may be an area in which standardized testing itself can be repurposed. Imagine the difference between the following two curriculums:

(1) The objective of the learning is to perform well on a standardized test at the end of the year. As with so many other activities, the emphasis

on the end goal can actually inhibit the types of interdisciplinary, broad learning so valuable in a world of increasing complexity. This is because the standardized tests are, within a sometimes relatively small amount of variability, a "known" commodity. Teachers whose performance evaluation – and thus remuneration and opportunities for advancement and increased autonomy – is dependent on the performance of the students on these standardized exams are actually encouraged to throttle learning at some point in order to ensure requisite performance on the standardized tests. The relationships between different fields of study are often (but admittedly not always) de-emphasized in an effort to focus on the specifics of each of the subjects that will be tested. Without proper care, this method can actually reduce the amount of learning that occurs.

(2) The "standardized" testing is focused on real-time assessment of the student's ability to ingest the current course material, with the purpose of making curriculum-guided but personalized decisions for the content to be learned. The assessment is anticipated to be a tool to help the educator with assessment but not to replace the educator. Rather than limiting the educator, this form of ongoing analysis – heavily banked on text analytics – can provide suggestions for the content, the media, the learning philosophies, and the perspectives on the individual student's education. For example, video, in addition to text, has been shown to aid cognitive and metacognitive processes in some forms of training [Bals05], but the role of video in learning varies based on the student's personality, proficiency, and preferences, among other factors.

Obviously, given the themes of this book, we argue for the second approach. However, a fair rebuttal is that, without reliable analytics, it is a lot more difficult to administer. One goal of the functional text analytics approaches outlined in this book is to help make the administration easier.

Our final chapter-opening quote comes from Abigail Adams, the calmer partner in one of early America's stronger families. Although she is not known to have been an enthusiast of text analytics, she still speaks volumes by noting that learning is not random and is achieved through careful application and monitoring. This supports a customized approach to education when the goal of education really is to promote learning as much as possible in the time allotted. It should be kept in mind that not all of the goals of the Common School Movement (Horace Mann being one of the key proponents) were educational; for example, the movement was also meant to make teaching a profession and to put education firmly under the control of the government [Chur76]. These "adjacencies" to education should not be forgotten in our zeal to create functional text analytics. For example, one important aspect of public education is to provide a place where a child from a disadvantaged background might receive one or more nutritious meals during the course of the day. An aching stomach due to hunger is certainly an impediment to learning, and bio-recording analytics may be able to assess a student's receptivity to learning based on blood glucose level, attentiveness [Liu17], and other measurements. But, we have not arrived at that day yet, and these factors are outside the scope of this text. What Abigail's quote directly supports in this book is the decision to allow the curriculum to adapt to the student and not just force the student to adapt to the curriculum. We move from a common school to one that is uncommon.

6.2.1 Metadata

Will Rogers might have said, "I never metadata I didn't like." Except, he did not. But, he did not say "I never met a man I didn't like," either, so let us not get touchy. The point is, **we** like metadata. Even when it is wrong, it can potentially tell us something about the information with which it is associated. Metadata is the context around the data itself and, as such, is an important source of information about the content. Familiar forms of metadata include tagging (or labeling), indexing, clustering, classification, and file metadata. Tagging (or labeling) is a descriptor or designation about what the content is or what it is about and includes file metadata such as file type, date created, date edited, creator, security policies, target audience, location, access rights, and keywords. Indexing is usually a narrower set of descriptors which provide the subjects (topics) of the content. Tagging is also referred to on occasion as cataloguing, and indexing is occasionally viewed as synonymous with categorizing. These definitions are less important than what they represent: some of the many varieties of text metadata.

Other important forms of text metadata are information derived from the content of the text itself, and it is here that we bridge from general analytics to text analytics. Keywords, for example, can be derived from purely statistical approaches – the keywords for a document being the words with the highest relative frequency compared to the other documents in the corpus, for example. There are alternatives to statistical derivation, however. For example, keywords can be entered by the creator or any recipient of the document. Keywords can, alternatively, be selected from a list of allowed or recommended labels. Keywords can also be generated using machine learning approaches; for example, by pooling documents that have been assigned to the same cluster or category and deriving a shared set of keywords for all of the documents in this group.

At a larger scale than keywords, a summary of the document is a form of metadata since it functions as a descriptor for the document much in the same manner as a keyword does. Word counts and dot products of word frequencies with representative documents of a cluster, class, or category of documents are other forms of "larger scale" metadata. Considering each of these as metadata makes sense once one considers it from a functional perspective. The function of metadata is largely to provide access points to the document. Whether the document is going to be used as the result of a search query or as an element in a sequence of documents used for an educational purpose, metadata creates access points for the document. Metadata is how data is made available to others.

6.2.2 Pathways of Learning

Earlier in this chapter, we addressed a simple means of determining a document reading order. We also showed its value for being able to assess the relative accuracy of multiple summarization engines. From a functional text analytics standpoint, however, pathways of learning can be viewed in a manner similar to the well-known "traveling salesperson" problem. In the traveling salesperson problem, the goal is to determine a pathway that visits each of several (perhaps many) locations only once but uses the minimum overall distance. However, the traveling salesperson is an oversimplified problem in that travel does not always permit one to visit each location only once. The US domestic air industry, for example, uses regional hubs to concentrate incoming and outgoing traffic in order to maximize the percent of capacity filled on flights. So, if you are using air travel to visit all of the locations, you might end up visiting Chicago or Atlanta several times to visit a required set of US cities. If you are using highways for your travel, then you must pass through Haines Junction, Alaska, twice to visit Haines, Alaska, and through Carcross, Yukon Territory, twice to visit Skagway, Alaska. Constraints such as these make the sequencing of information more problematic.

Pathway planning gets more difficult in the unpredictability of the real world. What if, halfway between Haines Junction and Haines, you find out that the road has closed? Would it make more sense for you to turn around and make the trip to Skagway and double back when the road reopens? Only if the road reopens a substantial number of hours later; otherwise, it makes more sense to wait it out on the Haines Highway. Worse yet, once you reach your destination, what if you find out the purpose of your visit cannot be fulfilled (e.g., the business is closed for the day)? Then, you have to revisit the same spot. Clearly, the life of a traveling salesperson is a lot messier than a simple optimization algorithm would have us believe.

Let us now apply these principles to the problem of "visiting" a set of training content in order to attain a specific proficiency in a topic. Like the salesperson, the student needs to visit every node of content.

Ideally, the travel between nodes is minimized; that is, text elements do not need to be read multiple times. However, when learning, sometimes the same material needs to be revisited. This can be done for reinforcement, for intense studying for certification or other examination, for refreshing, for continuing education, and for many other purposes. The point is that educational plans can get interrupted before they are complete, meaning the reading order sequence determined at the onset of the learning may need to have an open-ended design from the start. Such a design requires an open-ended evolutionary algorithm, which will support what we call a **conditional reading order**. This is considerably more complex than the reading order proposed above, exemplified by Equations (6.4) and (6.5). With a conditional reading order, multiple sequences are continually being evaluated as each document is being read. Figure 6.1 shows the system diagram for a conditional reading order process. The learner is assigned a Current Document based on their attributes, including assessed proficiency and the previous document read. Each time a new Current Document is delivered to the learner, assessment criteria relevant to the learning situation are created. These can include functional measurements of proficiency, such as the ability to perform specific tasks. One example of this is by displaying proper use of new terms and concepts in assignments or ongoing projects. Another is the time that passes before they complete the Current Document and request the next one.

Continuing with Figure 6.1, the assessment is performed. The output of the assessment is evaluated against the requirements for proceeding with the Next Best Sequence Candidate 1. This sequence candidate might require a proficiency score of 0.6 and an interval since the Current Document was assigned of less than 10 days. If 11 days have passed or proficiency is assessed at 0.57, then the assessment fails and the criteria for Next Best Sequence Candidate 2 are compared to the learner's assessment. This continues until a Next Best Sequence Candidate is assigned, at which point a new document is delivered, and it becomes the new Current Document.

The approach outlined in Figure 6.1 is adaptable. For example, Figure 6.1 implies that a Next Best Sequence is assigned that has multiple documents in a queue. Under that circumstance, it is anticipated that the learner will continue reading these documents in order until a given criterion is no longer met. This is a generalized approach which allows the ready incorporation of "chunks" of existing curricula. For example, if the learner is becoming proficient at human anatomy, such a "chunk" may include a sequence of five materials dealing with the bones, tendons, ligaments, muscles, and connective tissues of the hand. Since these documents are learned together and the curriculum for it does not require (or is not capable of) variation, the entire chunk is treated as the Current Document, and the Next Best Sequence

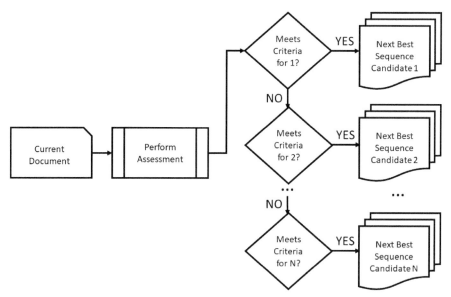

Figure 6.1   Analytics system diagram for conditional reading order. Each time a new Current Document is assigned, the assessment criteria are collected. Once the document is completed, the assessment is administered and collected, if applicable, and then applied to each of the multiple Next Best Sequence Candidates of content. It is anticipated that the individual will proceed learning with the given sequence until its criteria is no longer met, although it is possible that a sequence of length 1 is always computed, and so the next best document is reassessed in real time upon completion of the Current Document. Please see text for further details.

Candidates are not assessed until the learner has completed all five entries in this aggregate Current Document. On the other extreme, it is entirely feasible that each Next Best Sequence is a sequence of length 1. Thus, the ND is always decided upon when the Current Document is completed. This is indeed a very flexible system.

6.3 Machine Learning Aspects

Aside from the obvious shared term, there is a lot of room for machine learning in the field of learning. Machine learning has long been driven at the architectural level by investigations into how intelligence is architected in nature. From genetic algorithms, which are based on the way eukaryotic cells reproduce, to neural networks, which are based on the way multiple-layered cortices in brains process information, machine learning is concerned with allowing enough variability at each step to provide robustness while providing enough feedback and reinforcement to provide convergence onto highly accurate behavior. As more has become known about how biological systems, such as the visual and immune systems, work, this learning has been internalized into the field of machine learning. Convolutional neural networks, for example, are analogous to the type of lateral inhibition and edge enhancement algorithms that occur in the retina and in the layers of the occipital cortex. Swarm intelligence is based on certain aspects of the immune system and, in particular, the behavior of leukocytes.

Training as the efficient delivery of sequential learning material is an important concept, with the learner's current state and aspects of the learner's past state being the key input to determine the future state. This is directly analogous to Markov chains, in which the most recent states have the strongest influence on the next state. One area for consideration for deeper research in the field of reading order optimization is most certainly that of Markov models.

6.3.1 Learning About Machine Learning

You may be thinking, "But, this is not machine learning — it is the learning of material." And, you would be right. But, since machine learning is a product of human intelligence (not to mention often being modeled on human intelligence), it behooves us to close the loop and have the machine learning provide some suggestions for enhancing human learning. Some of the more interesting machine learning approaches which, at least on the surface, seem to diverge significantly from human learning are massive Bayesian networks, multi-factor or hybrid machine learning approaches, and evolutionary algorithms. Humans are notoriously poor at statistics (even many engineers who are gifted at electromagnetic field or tensor theory), and Bayesian networks often work counter-intuitively. Handling many degrees of uncertainty at once is more the role of the autonomous nervous system than that of the conscious brain. One way in which Bayesian networks can be made more accessible is for the rules that they produce to be stated in normal vocabulary; that is, reduced to "expert rules" that can be deployed. The same can be said of multi-factor or hybrid algorithms. Evolutionary and immunologically inspired algorithms, on the other hand, are likely to continue having an "emergent" behavior that somewhat defies easy explanation.

Of the many possible future threads for hybrid algorithms, two stand out for mention here. The first comes from the field of meta-algorithmics [Sims13] and is the design pattern for intelligent systems called predictive selection. In this design pattern, an upfront set of analytics are performed, not with the view of analyzing the text information but, instead, for generating a set of **predictive** values used for **selecting** the downstream analysis to best provide the desired accuracy, robustness, performance, or other objective function outputs of significance (cost, resilience, ability to process in parallel, etc.). Metadata is an excellent predictive value in many cases; for example, the author of the document, the original language in which the document was written, or the revision number of the document may all provide differentiation for the text analyst. Documents written by William Shakespeare, Edgar Allan Poe, and Anna Burns, though all composed in English, may require largely different text analytical approaches due to differences in vocabulary, poetic content, and surrealism, to name just a few factors. In the case of drafts, there will typically be a different, perhaps more polished, set of expressions as the draft number rises. Regardless, simply assigning inputs to separate categories based on attributes of the documents affords the opportunity for distinct, parallel pathways for downstream analysis.

The second potentially valuable hybrid approach is the combination of Bayesian and evolutionary algorithms. As mentioned above, both of these have largely different mechanisms for learning than do, for example, neural networks. Combined, they hold promise for potentially useful new means of learning that allow the benefits of both of these approaches to be felt. For example, Bayesian methods can be used to determine the settings for the evolutionary algorithm (e.g., mutation rate, crossover rate, survival rate, etc.), and the evolutionary algorithms are allowed to change the probabilities associated with the Bayesian network. In this way, the network iterates over time, eventually stabilizing on an optimized design for the salient training purposes (content selection, content sequencing, content testing, etc.).

6.3.2 Machine Learning Constraints

Think, if you will, of the way in which children are taught about the world. Once children reach the age where they become aware of the world, they move on quickly to trying to understand it. What are its rules, its limits, and its expectations of the child? Good educators, including their parents, will provide the appropriate structure for the child to learn and develop. A common pithy expression is that without constraints, children become savages and that with too many constraints, they may become uncreative and timid. This is undoubtedly an oversimplification, but the point is well-taken: learning is a guided process. With too much structure, it can be lifeless and tedious. With no structure, it is more like exploring than learning. Certainly, each of these two extremes work for the rare individual who wants to be hand-held every step of the way or the other rare individual who never wants anyone offering a suggestion. For the rest of us – all of us humans – some structure helps, if for nothing else than framing the discipline we are studying and providing a summary of the current state of the knowledge.

Machine learning is no different – structure for how machines learn is just as important as the choice of the algorithm. Bad training results in bad performance when deployed. This is because machine learning, like human learning, can set poor habits in place without the means for remediation. We need to have deep unlearning as much as deep learning if we are to work with machine learning the same way we do with our human colleagues. Machine learning habituation (learning to forget) is important. If you have ever met someone with comprehensively photographic memory, it is a bit unnerving that they remember every word spoken in a conversation from a decade ago. One constraint for machine learning is that, without reinforcement, certain connections should be allowed to fade over time. This can be accomplished simply by updating the input data on a schedule.

What about machine teaching, machine studying, machine rehearsing, and even machine certification? Clearly, we will have achieved something fundamental in machine learning when each of these aspects of the human learning (and unlearning) process can be replicated in code. Machine teaching can be made more natural when, for example, the machine learning is based on an ensemble. Even ensembles of neural networks can be used. Trained on different subsets of the input, these networks can be emphasized or de-emphasized over time as the input of the system changes. An example of this is shown in Table 6.3, in which case the input is assigned to three different groups, and each of the three neural networks is trained on two of the three input sets. When the balance of the three types of input ($A = 33.3\%$, $B = 33.3\%$, $C = 33.3\%$) changes to A-emphasized ($A = 50\%$, $B = 25\%$, $C = 25\%$), B-emphasized ($A = 25\%$, $B = 50\%$, $C = 25\%$), or C-emphasized ($A = 25\%$, $B = 25\%$, $C = 50\%$), the optimal combination of neural networks to decide on the output also changes. Note that these percentages are changed by simply doubling the input value of one of the three types of neural network input. In the example, the accuracy of the three emphasized systems actually increases due to the ability of the combined networks to specialize on the inputs.

The results of Table 6.3 can also be viewed as a form of machine rehearsal, wherein the "emphasis" allows each of the three networks to "rehearse" what they know about 2/3 of the input. An alternate combination of the three neural networks is presented in Table 6.4. Here, the individual neural networks are trained on 1/3, instead of 2/3, of the input. The effect of emphasizing one input versus the other has negligible effect on accuracy.

Table 6.3 Machine teaching using three neural networks, each trained on 2/3 of the training data. The training data comprises thirds *A*, *B*, *C*. After the system is deployed, *A*, *B*, or *C* input is emphasized (becomes the predominant input, from 33% to 50%). The effect of the change in input shows the changed optimal weighting of the combination of the three networks. Some improvement in overall system accuracy is observed.

Neural network	Original weights $(A = B = C)$	Weights, *A* emphasized	Weights, *B* emphasized	Weights, *C* emphasized
1 (*A*, *B* trained)	0.333	0.407	0.377	0.226
2 (*A*, *C* trained)	0.335	0.399	0.208	0.405
3 (*B*, *C* trained)	0.332	0.194	0.415	0.369
Accuracy	**0.885**	**0.892**	**0.899**	**0.895**

Table 6.4 Machine teaching using three neural networks, each trained on 1/3 of the training data. The training data comprises thirds *A*, *B*, *C*. After the system is deployed, *A*, *B*, or *C* input is emphasized (becomes the predominant input, from 33% to 50%). The effect of the change in input shows in the changed optimal weighting of the combination of the three networks. The overall increase in accuracy is negligible here.

Neural network	Original weights $(A = B = C)$	Weights, *A* emphasized	Weights, *B* emphasized	Weights, *C* emphasized
1 (*A* trained)	0.333	0.517	0.261	0.271
2 (*B* trained)	0.335	0.235	0.496	0.265
3 (*C* trained)	0.332	0.248	0.243	0.464
Accuracy	**0.885**	**0.887**	**0.884**	**0.889**

The differential results of Table 6.4 may illustrate an analogy to human learning, as well. Better long-term memory and mental performance is obtained when more of the brain is involved in the memory. In Table 6.3, 2/3 of the neural networks are involved in the decisions for each type of input; in Table 6.4, only 1/3 are. Thus, the decisions made on the emphasized inputs for Table 6.3 involve twice as much neural network involvement as for Table 6.4. Better memory involves more of the brain?

6.4 Design/System Considerations

In some ways, the design and system considerations for learning can be summed up as the syllabus or curriculum: the plan of training along with its specific elements. The sequencing of training information and the manner in which information is archived, accessed, and made available for repurposing are all – to varying extents depending on the type of document – dependent on the overall training goals.

6.4.1 Do Not Use Machine Learning for the Sake of Using Machine Learning

There is always a temptation to use a new tool when you first get it. The cliché is "when you have a hammer, every problem looks like a nail." Right now, every problem looks like it needs machine learning. But, machine learning has limitations. In general, a lot of data is needed to train machine learning algorithms. This means a lot (more) data is needed to un-train the machine learning. We are currently in a time in the history of algorithms where many of us (ourselves included) are carefully crafting situations in which the machine learning that we are fond of will perform better than the alternative. We need to allow the situation to dictate the technology employed, and not the converse.

The implications of this on machine learning test and measurement are not to be underestimated. A separation of powers needs to be enforced during the creation of test and measurement experiments for machine learning based text analytics, along the lines of the familiar "red team, blue team" roles in cyber security. The person(s) responsible for generating the training data need to be different from those who create the algorithms. Otherwise, not only will the system be over-trained, it will also be over-promised. Would you not rather deliver on what you advertise? The single best way to be able to predict how well a system will perform in the real world is to simulate real-world conditions as much as possible during the training and validation stages.

6.4.2 Learning to Learn

Humans do not go to school to learn everything. Aside from the oxymoron of "learning everything," who would want their mind filled with all of the mistakes of the past? Who would want the orthodoxies of 100 years ago (think of Freud, phrenologists, and phlogistons, just to name an alliterative triad of folly) to occupy their consciousness? The purpose of education is to learn how to learn. As Mark Twain noted, "When I was a boy of 14, my father was so ignorant I could hardly stand to have the old man around. But when I got to be 21, I was astonished at how much the old man had learned in seven years." Mark had learned how to learn. This seems natural in humans, and for all but the most obstreperous or ornery of intellects, we humans are generally excellent at being able to place more and more in some form of relevant context as we mature.

We need to be able to do the same for machine learning. Enhancing, understanding (placing in context), and selective retention of what is most important can all be addressed if the machine learning process includes room for changes in weighting. In other words, machine learning will be more adaptable either when it provides an ensemble of intelligent algorithms or a real-time adjustment to a specific algorithm. A key design consideration for machine learning is to be open to changes in input. This is one advantage of statistical approaches, such as those highlighted in the extractive summarizations of Chapter 2. Statistics are readily updated with new data, and the output is directly derived from the statistics. This means such statistical approaches are real-time adjustable to input changes.

6.4.3 Prediction Time

Another important element of the design of intelligent text analytics is the temporal representation of a corpus. As the Wayback Machine reminds us, the Internet changes almost beyond recognition in a short period of time. It is highly likely that the same is true of any data set of relevance to your business, your research, or your finances. Machine learning algorithms should preserve their time-stamped versions rather than simply being retrained. The worst-case scenario is that you have preserved a system that would not get used often; however, when it is needed, what an advantage you will have. One of the great dangers of any time in history is judging the past by the lenses of today. Future generations will look back on us and condemn us for our barbarism. What will we be considered barbaric about? Some of it we can see (indigence, racism, sexism, and the like). But, other elements we are likely oblivious to, and we will be condemned for it. When people use a revisionist outlook on people of the past, there is always a hateful flaw that we can find. This is because somebody forgot to store the algorithm to properly interpret the past. We should not make that mistake. In order to truly understand the benefits and limitations of machine learning, we must preserve how it changes over time, not just allow it to change over time.

6.5 Applications/Examples

Applications of sequential flow of text analytics including reading order optimization, curriculum development, customized education planning, and personalized rehearsing. Reading order optimization was covered earlier in this chapter; here, we discuss the additional applications.

6.5.1 Curriculum Development

What could be more germane to learning than the development of a robust, effective curriculum? With this in place, learning is guaranteed, right? Well, not so fast, because one very important aspect of learning is the messenger.

Training is delivered in a much different fashion now than it was 100 years ago. With the continuing increase in video lecturing along with the advent of AR/virtual reality (VR), it is clear that text-based learning is a teammate in the future of learning, and not the entirety of the learning material.

The functional aspect of customized learning is addressed through assessment of improved proficiency. The functional aspect of curriculum development, however, requires the assessment of the entire complement of students to whom the text materials are being delivered. There are several strategies for the assessment, including the following:

(a) Performance of the best student/student cadre (performance meaning score on a suitable assessment element);
(b) Performance of the worst student/student cadre;
(c) Median performance of the student cadre;
(d) Improvement in the median student score after delivery of the learning content;
(e) Improvement in the rate at which students digest (that is, read) new material;
(f) Self-assessment by the students.

Of course, some combination of all of these factors is more holistic and, likely, a more accurate assessment in most cases. One other factor of note here is, of course, cost. We have been relatively quiet on this factor, but any time an objective function is used, cost is logically one of the elements. In real-word situations, the content for providing up-to-the-minute educational materials has expense involved. The more valuable the content, typically, the more skilled the creator of the content. We are not fans of pirating because it discourages content creators and the differential talents of those highly skilled in areas of interest to learners. So, in addition to the six factors above, content accessing costs are important. They can typically be justified on the differential career earnings that the educational program helps bring.

6.5.2 Customized Education Planning

The topic of text sequencing for general-purpose text content delivery was originally discussed in Chapter 1. There, its value in a learning, training, retraining, or continuing educational environment was emphasized. The timeliness of delivery of content is not just significant in a sequence, but as it relates to the receptivity of the reader in general. At its most comprehensive, reading order assesses your entire history of reading and determines the best document for you to read next. This document should account for your current motivation, gaps in knowledge and understanding, and trajectory of your specific learning goals. You are the demand side for this content. On the supply side is the context provider, who is concerned with the balancing act of providing you with sufficiently new content to promote learning, but sufficient overlap with your current set of knowledge to enhance receptivity to the content and understanding. We know that receptivity correlates strongly with the speed of learning. We can decide to use minimal overlap with previous material of the "suggested next document to read" for the fastest learners or those most proficient with the material; intermediate overlap with previous material of the "suggested next document to read" for the intermediate learners or those moderately proficient with the material; and maximum overlap with previous material of the "suggested next document to read" for the slowest learners or those least proficient with the material.

In addition to reader proficiency, we are concerned with the purpose of the learning. For example, the reader may be undergoing continuing education, remedial learning, or intensive studying in preparation for a certification examination. In such a situation, a learner of moderate or even expert proficiency may prefer a "lower proficiency" setting for the purpose of more deeply studying and/or memorizing the material. This illustrates the interaction between curriculum and purpose. We now turn to the actual mechanism for the learning; that is, the timing of the content delivery.

6.5.3 Personalized Rehearsing

When rehearsing material, the best next text content (ouch, do not say that out loud!) to be delivered is whatever content you need to set into memory. If you are playing a part in a community dinner theater, for example, the content you need are the lines you just cannot seem to commit to memory. It is not enough to simply assess the proficiency of the learner; you must also provide understanding of their needs. This is truly functional learning: the content delivered to you is precisely what you need at the moment. In addition, the delivery process must be able to (1) verify that you received the content; (2) you were receptive to the content; and (3) you acted on the content.

It is not far-fetched to imagine a world in which redundant machine intelligence systems are monitoring you at all times. The overlap between these systems provides consensus using ensemble and weighted voting assessments, and, as a consequence, the meta-systems incorporating input from all of the systems currently tracking, monitoring, and assessing you can predict with increasing precision what your next content need is going to be. Personalized rehearsing, in this context, can provide you not just with what learning you need next but with what coaching you need next. As more of our society and culture moves from the physical to the cyber realm, it may well change what the very concept of learning is. For example, a key learning for most people these days is how to use online search. Without these (perhaps) simple but extremely valuable skills, you are missing the library with the greatest amount of information possible. This is meta-learning or, from a certain perspective, another form of "learning how to learn." It is inevitable with the Internet, mobile devices, AR, and intelligent digital assistants for learning to become more about knowing how to find information than how to memorize it. Maybe with the exception of passwords, but then they may be the exceptions that prove the rule. In the future, personalized rehearsing may focus on reminding you how to remember passwords and to remember how to access everything you "know" online. It is a brave new world, indeed.

6.6 Test and Configuration

Testing of a learning plan is integrally related to the development of a curriculum. Just as a course depends on its syllabus, a learning plan depends on its curriculum. In some ways, the curriculum defines the boundaries of knowledge for a given cadre of students. The intent of advanced educational planners, however, is to make sure that these boundaries are Schengen boundaries, with the students freely able to move across them as needed for their personalized and dynamic learning plans. To any modern educator, learning is no longer a linear path from ignorance to proficiency. Instead, learning is a (virtuous) circle comprising these steps of test, measure, assess, retain, and repeat as necessary.

In this chapter and book, we have focused on the need to generate functional assessment strategies. The nature of information, and the speed at which topical knowledge advances, is simply too much for traditional assessment to be the only means of determining proficiency. Machine learning approaches are assessed by their relative capability to solve specific tasks. Accuracy is simply a measurement best used for the settings, and not as the end goal. Measurement, where possible, often makes more sense than testing. If you read twice as fast as you used to and integrate new vocabulary into your own speech and writing with proper usage as part of your job, then you do not need separate assessment. Functional assessment normally suffices. As we all know, it is easy to tell when someone is good at their job. We do not need multiple choice, true/false, or short answer to grade them: we can see it in how they *function*.

Assessment, then, plays a more specialized role in this educational system. We assess to pass an audit. We assess our teachers to ensure that they can sufficiently explain the likely conundrums of their students. And we assess to make sure that the content in the assessment is still relevant. Assessment may be properly concerned with determining if the right information is retained. Twenty years ago, knowing how to buy books, shoes, and pet supplies online required remembering a list of online retailers. Now, many people select a one-stop shopping experience, and what they need to retain is more about the pricing discounts and bundles from that one provider. Nowadays, meta-information changes as quickly as the information it aggregates. We are moving one level of indirection away from content. Learning will thus increasingly become about meta-content. This, in turn, leads us to change the way we test, measure, and assess. And so, the circle is completed.

6.7 Summary

This chapter describes learning as the culmination of applications for functional text analytics. Learning is viewed as a sequential delivery of content for the purpose of training. While learning, the role of functional text analytics is to determine what the "next best document" to deliver to the learning is. This is based on their past proficiency as well as their ongoing proficiency with the current document. A functional means of assessing the effectiveness of a summarization engine is shown to be comparing the outputs of a reading order algorithm for both the summaries and the original documents. The reading order of documents to read can depend on a number of factors, including the reader's proficiency, the relationship between the documents, and the amount of time the reader has to dedicate to the task. Next, vocabulary proficiency as a functional means to assess learning efficiency was described. Reasonable purposes for standardized testing, for applying metadata to documents, and for determining conditional reading orders were then presented. In the machine learning section, some analogies between machine learning and human learning are explored, and the possibilities of the two fields helping to advance each other. In particular, the analogy between employing ensembles of neural networks and rehearsal in humans was proposed. The system and design concerns for learning are shown to highly depend on the training goals of the reader. Personalization of learning, both in the individual and in the entire cadre of students, was considered next. Here, we noted the blurring of the lines between learning and learning how to find learning online, which is certain to continue changing the nature of education in years to come. Learning as a cycle of test, measure, assess, retain, and repeat completed the chapter.

References

[Alha18] Alharthi H, Inkpen D, Szpakowicz S, "A Survey of Book Recommender Systems," Journal of Intelligent Information Systems, 51(1), pp. 139-160, 2018.

[Bals05] Balslev T, De Grave WS, Muijtjens AMM, Scherpbier AJJA, "Comparison of Text and Video Cases in a Postgraduate Problem-Based Learning Format," Med Educ 39(11), pp. 1086-1092 (2005).

[Beel16] Beel J, Gipp B, Langer S, Breitinger C, "Research-Paper Recommender Systems: a Literature Survey," International Journal on Digital Libraries, 17(4), pp. 305-338, 2016.

[Chur76] Church R, "Education in the United States, an interpretive history," New York: The Free Press, 489 pp. (1976).

[Day84] Day WHE, Edelsbrunner H, "Efficient Algorithms for Agglomerative Hierarchical Clustering Methods," Journal of Classification 1, pp. 7-24 (1984).

[Fors09] Forshaw RJ, "The Practice of Dentistry in Ancient Egypt, British Dental Journal 206, pp. 481-486 (2009).

[Khu16] Khusro S, Ali Z, Ullah I, "Recommender Systems: Issues, Challenges, and Research Opportunities," Information Science and Applications (ICISA), pp. 1179-1189, Springer, Singapore, 2016.

[Liu17] Liu X, Tan PN, Liu L, Simske SJ, "Automated Classification of EEG Signals for Predicting Students' Cognitive State during Learning," ACM WI '17, Leipzig, Germany, pp. 442-450 (2017).

[Oliv18] Oliveira H, Lins RD, Lima R, Freitas F, Simske SJ, "A Concept-Based ILP Approach for Multi-Document Summarization Exploring Centrality and Position," 7th Brazilian Conference on Intelligent Systems (BRACIS), 2018.

[Sims13] Simske S, "Meta-Algorithmics: Patterns for Robust, Low-Cost, High-Quality Systems", Singapore, IEEE Press and Wiley, 2013.

[Sims19] Simske S, "Meta-Analytics: Consensus Approaches and System Patterns for Data Analysis," Elsevier, Morgan Kaufmann, Burlington, MA, 2019.

[Tar18] Tarus JK, Niu Z, Mustafa G, "Knowledge-based recommendation: a review of ontology-based recommender systems for e-learning," Artificial intelligence Review, 50(1), pp. 21-48, 2018.

7

Testing and Configuration

"Testing oneself is best when done alone"
– Jimmy Carter

"I have to keep testing myself"
– Eartha Kitt

"The future's uncertain but the end is always near"
– Jim Morrison

Abstract

Testing and configuration are two of the important systems engineering aspects of text analytics, and they provide the framework for how the text analytics are applied. Testing is not viewed in isolation of the text analytics system but is, instead, built in with the analytics from the ground up. This Data-Ops approach is consistent with the concept of operations (CONOPS) and security + operations (SECOPS) approaches used in systems engineering. In this chapter, the advantages of using a distinct analytical process to perform testing and configuration validation on another analytics process are shown. This approach, a functional approach to text analytics testing, is shown to have advantages in building more robust text analytics overall.

7.1 Introduction

In the first six chapters of this book, we have presented a multiplicity of ways in which functional text analytics can be used to simplify the training, validation, and evaluation of text-related intelligent systems. In this chapter, we illustrate the advantages of functional approaches further, showing their utility both for the testing and measurement of deployed systems and for ensuring that the configuration (design, architecture, and settings) of the deployed system are functionally optimized for their intended tasks.

Our premise in applying functional text analytics is that overall system outputs can be substantially improved when a distinct text analytic process B is employed to provide test and measurement capabilities for the specific text process A for which we are interested in providing the best possible output. In other words, when measurement of an adjacent text process is most positively (or least negatively) affected by employing the specific process that we wish to optimize, we have determined our optimum system. Rather than relying on the training or validation data for process A to determine the optimum configuration of process A, we employ the various configurations of process A and find out which one of those results in the best outcome for process B. This is advantageous to us for several reasons, including:

(1) Ground truth (human-entered training) is either not required or it is moved to an application that is easier to train (i.e., ground truthing takes less effort). Since human ground truthing is expensive, both time-wise and potentially incentive-wise, any process by which the ground truthing process can be simplified is generally advantageous.

(2) By using a second process to determine the settings for the first process, a more robust system is likely created. This is analogous to iterative algorithms such as expectation–maximization and the more general competition–cooperation as described in earlier chapters. In these approaches, two different means of organizing content are played off each other in each iteration, usually resulting in some form of system stability after a given number of iterations. At the macroscopic (system) level, using one system of outputs (e.g., the behavior of a corpus in response to a set of search queries) to gauge the optimum settings for another system of outputs (e.g., the best summarization approach from among a set of possible summarizations) generally provides for an overall more stable system, particularly when the two systems of outputs are not highly correlated. The system stability generally results in the system being more robust to new input, as discussed in several examples in earlier chapters.

(3) Of the most interest to the goals of this chapter, this dual-system approach results in an overall more comprehensive testing of the capabilities of the individual system. As an example outside of the text analytics domain (to show the generality of the approach), suppose that you are deploying a system to track individuals with multiple cameras and wish to optimize the settings of the cameras to provide peak

authentication accuracy. Actually providing training sets for tracking and measuring the accuracy is time-intensive and expensive. Instead, a functional means of assessing the tracking accuracy is to have each of the cameras count specific items as they pass a region of overlap. For example, suppose there are four cameras, pointing north, south, east, and west at a particular intersection. A simple vehicle counting algorithm can be used to optimize the overall system, as shown in Table 7.1. Here, the video camera settings of the greatest outlier camera are changed iteratively until the four cameras agree on vehicle count through the intersection. In the last column of Table 7.1, the tracking accuracy (performed here only to show the actual improvement in the system) is given. The accuracy improves as the vehicle counts for the four cameras become more equal. After six adjustments, all four cameras count the same number of vehicles, and tracking accuracy has improved 0.847 − 0.675 = 0.172, or 17.2% over the default. In a separate tracking optimization approach (not using the vehicle count but, instead, using direct training data for the optimization), a slightly higher accuracy of 0.855 (an improvement of only 0.8%) was obtained.

In the example of Table 7.1, it is clear that the closer the vehicle counts get to each other, the higher the tracking accuracy. Plotting the largest (absolute value of) difference in vehicle count versus the tracking accuracy yields the plot given in Figure 7.1. The correlation coefficient, R^2, is very high for the relationship between these two factors (0.955). In general, the higher the

Table 7.1 Tracking system accuracy improved by adjustments of the settings on the cameras. The left column lists the seven consecutive configurations used. The first was with default settings for all four cameras, and then the camera furthest from the others in terms of vehicle count was adjusted (North). This process was continued until the four vehicle counts (off of the video from all four cameras) were in agreement. The rightmost column shows the continued improvement of the tracking accuracy of the system by using the vehicle count as a substitute for tracking ground truthing. In a separate study, when tracking accuracy was optimized using ground truth data, a slightly higher 0.855 accuracy was obtained, but this is only 0.008 more than the 0.847 obtained using the vehicle count matching approach.

Configuration	Vehicles (North)	Vehicles (South)	Vehicles (East)	Vehicles (West)	Tracking accuracy
1 (Default)	13	21	19	20	0.675
2 (North)	16	22	20	21	0.744
3 (North)	19	20	19	22	0.795
4 (West)	21	20	21	20	0.822
5 (South)	22	23	22	21	0.835
6 (West)	24	23	24	24	0.841
7 (South)	22	22	22	22	0.847

correlation coefficient between the functional metric (here vehicle count) and the desired metric to optimize (here, tracking accuracy), the better the functional metric is able to serve as a proxy for the otherwise expensive training of the system.

In general, the higher the correlation coefficient, the closer the results will be for the desired system output when the functional system optimization is complete. In the example here, the improvement is 0.172/0.180, or 0.956, as much. This value is not surprising since its expected value is the correlation coefficient (which it just happened to be equal to here).

These results highlight several important systems aspects of the functional approach to analytics (in which a different output is used as a proxy for optimizing the settings of the desired system output), which are captured here.

(1) Of the variety of functional approaches for system optimization at the avail, select the one that has the highest correlation coefficient with the actual system output to be optimized. For example, suppose that we are interested in optimizing document translation accuracy. We measure the behavior of the translated documents with translated query sets and find this correlates with the behavior of the untranslated documents with the untranslated query sets with an $R^2 = 0.964$. Next, we measure the correlation between the TF*IDF histograms (term frequency times

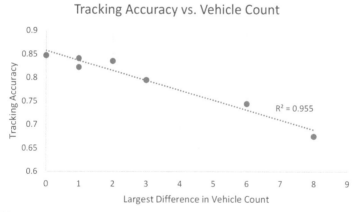

Figure 7.1   Plot of the tracking accuracy (column 6 of Table 7.1) versus the largest absolute value of the difference in vehicle count between any pair of cameras in Table 7.1 (largest absolute value of differences comparing any two of columns 2−5). There is a high correlation between the two measures ($R^2 = 0.955$), indicating that the vehicle count discrepancy is a strong negative predictor of tracking accuracy. This means that the smaller the discrepancy, the higher the tracking accuracy. Please see text for more details.

inverse of document frequency, please see Chapter 5) [Spär72] before and after translation and compute an $R^2 = 0.971$. Based on these two values, we determine that using the TF*IDF histograms provides a better proxy for document translation accuracy than does the query set behavior.

(2) The residual improvement of the actual system output has an expected value of 1.0 minus the correlation coefficient, R^2, as computed in (1). Thus, for the query behavior proxy, a residual improvement of $1 - 0.964 = 0.036$ is anticipated. For the TF*IDF histogram proxy, an improvement of $1 - 0.971 = 0.029$ is anticipated.

(3) Next, the cost of the residual improvement (in terms of the higher expense of ground truthing) is compared to the costs incurred by the higher error rate. If the cost of the incremental improvement in accuracy — for example, the 0.036 or 0.029 in (2) — is higher than the value of the incremental improvement in accuracy to the owner of the system, then the system designer is better off using the proxy. For example, if the cost of having 10% errors in document classification is $100,000, then the value of the increment accuracies of 0.036 and 0.029 are $36,000 and $29,000. If the cost of experimental design, performing the experiments, and paying human experts to ground truth the outputs of the experiments is $40,000, then we are better off using either proxy method to optimize the document classifier. If the cost of ground truthing is only $20,000, however, then we are better off using the ground truth to directly optimize the document classification system.

Having seen how a functional approach to text analytics can be employed, let us return to placing this in context of testing and configuration. We have previously discussed various uses of a compressed proxy (summarization) representation of a document in place of the original document set. One concern around using such a compressed proxy set is that training, validation, and optimization are based on a quasi-experimental approach, rather than a proper experimental approach. A quasi-experiment is one that is empirical; that is, the control and experimental group assignment are *a posteriori* so that it is used to estimate the causal impact of an experimental factor but without random assignment. This means such an experiment is almost certain to be impacted by one or more confounding factors. This is almost certainly the case even for the strong correlation between tracking accuracy and vehicle count in Figure 7.1. For example, both tracking accuracy and vehicle count agreement may both be highly correlated and dependent on the accuracy of the segmentation algorithm (making the segmentation algorithm the confounding factor).

In a large number of experimental conditions, it may not be feasible, ethical, or rational to perform the most direct experiment. Let us consider this from the perspective of experimental ethicality. There are usually ethical and/or structural exigencies which prevent a proper experimental design in many psychology-related research areas. One example is to test the impact of tobacco smoking on the development of another physiological disorder such as lung cancer. It would be unethical to randomly assign participants in the experiment to either the control or one of the experimental (e.g., 0.5, 1, or 2 packs day) groups (also known as test groups) since the assignment to an experimental group would have considerable health risks (not to mention it may be difficult to enforce). Thus, for this particular experiment, the assignments are after the fact, and the experiment designer does not have the ability or choice to change the independent variable. To this end, if only an experiment designed along these lines was available, a cigarette manufacturer might argue that there is a predisposition of people to smoke who already have a higher than average genetic risk of lung cancer. Some might argue that this is statistical apologetics, giving the cigarette manufacturers a loophole. It is not—we will close that loophole shortly. However, from the standpoint of a quasi-experiment, it can reasonably be argued that a physiological lung defect such as weakened alveolar linings, dilated bronchi, etc., may make a person more likely to smoke: the smoking might constrict the bronchi to normal levels and so alleviate discomfort. From an ethical standpoint, the onus should be on the cigarette manufacturers to establish that constricted bronchi lead to a higher lung cancer rate, but that only delays the direct experiment through the creation of another confounding factor to keep the argument going. Fortunately, there are means of establishing a proper experiment *a posteriori*. Here is where identical (that is, monozygotic) twins are an absolute boon to psychological research. Because they are born with equivalent genetic information, if we find identical twins with different smoking behavior, we can act as if they were assigned to these different groups *a priori* (since from the genetic standpoint their assignment is random). The existence of identical twins is used to elevate the quasi-experiment to an experiment.

Applying these concerns about experiments versus quasi-experiments to the use of the summary as a proxy for the original document, we need to make sure that the summarization data (which is only quasi-experimental) can substitute for the full-text documents (which, with proper definition of training, validation, and testing sets, provide an experimental data set). To this end, proper equivalency comparisons (e.g., similarity of search query behavior on the original and compressed proxy documents, or similar suggested reading orders on the original and compressed proxy documents, both as described in Chapter 2) are essential.

7.2 General Considerations

7.2.1 Data-Ops

Data-Ops, or operations based on data analytics, are an important part of functional text analytics. Ultimately, the purpose of text analytics is to improve the utility of the text in one or more applications or services. To this purpose, the importance of Data-Ops is to provide a map (usually a table or matrix, to be precise) between the needs of the data consumers and the functionality of the data itself. The state of the data serves as the nodes in the system, while the transformation(s) necessary to move the data from one state to the next are the edges. The Data-Ops plan, therefore, can take advantage of generalized graph theory, using nodes and the edges between them to define pathways. As an example of Data-Ops graphing, suppose that we wish to translate Russian speech into French text. We have text analytics software capable of translating Russian speech into (a) Russian text, (b) English speech, or (c) French speech. We include the translation into English speech as a possible intermediary (and, yes, extra) step based on the findings of Chapter 4, wherein it was found that English often served as the "central" language for the translation engine. This centrality of English means that there is often both non-uniformity and asymmetry of translation accuracy in a multi-language system. Once English speech is obtained, it can be transformed into English text with very high accuracy. Russian text can separately be translated into English text or French text. The English text can be translated into the final form, French text. French speech can also be directly transformed into French text. Piecing all of these data operations together, we arrive at the Data-Ops graph of Figure 7.2.

In Figure 7.2, a generalized graph consists of nodes that correspond to one of the three languages {Russian, English, French} = {R, E, F} and one of two data types {Speech, Text} = {S, T}. Thus, there are six nodes in Figure 7.2. The edges correspond to the transformations and translations (combined, these are designated "operations") mentioned before, and each of the edges can either change R, E, or F to another language or change S or T to T or S, respectively. A path in Figure 7.2, therefore, is a set of two or more consecutively traveled edges that convert RS to FT. Five different reasonable paths can be traveled, and they are the basis of the edge directions in Figure 7.2. These five are:

(1) Russian speech → Russian text → French text, or RS-RT-FT

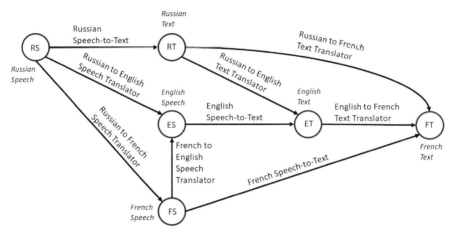

Figure 7.2   Generalized graph (nodes and edges) of a specific text analytics task, in this case, translating Russian speech into French text. Because an English speech and English text intermediate exists, there are five relevant paths for translation in the graph shown here. The uppermost path is from Russian speech to Russian text, and then to French text. The lowermost path if from Russian speech to French speech, and then to French text. Please see text for details on each of the five paths.

(2) Russian speech → Russian text → English text → French text, or RS-RT-ET-FT
(3) Russian speech → English speech → English text → French text, or RS-ES-ET-FT
(4) Russian speech → French speech → French text, or RS-FS-FT
(5) Russian speech → French speech → English speech → English text → French text, or RS-FS-ES-ET-FT

Next, accuracies (as probabilities of the correct result being obtained) are assigned to each edge in the graph. These accuracies come from the training and/or validation stages of the system qualification and are here used to select the best Data-Ops configuration. In Figure 7.3, these accuracy probabilities are placed on each edge, as follows:

(1) Russian speech transformed into Russian text, accuracy = 0.95
(2) Russian speech translated into English speech, accuracy = 0.88
(3) Russian speech translated into French speech, accuracy = 0.94
(4) Russian text translated into French text, accuracy = 0.94
(5) Russian text translated into English text, accuracy = 0.93
(6) French speech transformed into French text, accuracy = 0.93
(7) French speech translated into English speech, accuracy = 0.98

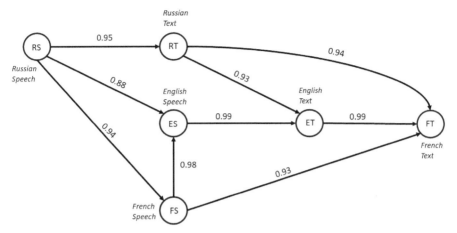

Figure 7.3 Generalized graph of Figure 7.2 with the accuracy (probabilities) indicated. If each edge in a sequence of edges traversing from RS to FT is independent of each other, then the accuracy (probability) of the path is simply all of the edge probabilities multiplied together. This assumption drives the values in Table 7.2.

(8) English speech transformed into English text, accuracy = 0.99
(9) English text translated into French text, accuracy = 0.99

From these accuracy values, we can see that certain operations in the overall system have extremely high accuracy; namely, any that transforms or translates from English. These high accuracies imply that English was likely the **language of centrality** for the system. Second, the accuracy of 0.93 for transformation (6), French speech to French text, shows that French was not likely a language of proficiency for the folks who built all of the different translation/transformation algorithms. Third, the accuracies of translating English speech into French speech, French text into English text, English speech into Russian speech, English text into Russian text, and French speech into Russian speech are not given. We therefore do not know how close these are to their inverse operations, which have accuracies of 0.98, 0.99. 0.88, 0.93, and 0.94, respectively. If they are significantly different from the accuracies of these inverse operations, then these are **asymmetric operations**.

Having added these accuracies (as probability values) to the edges of Figure 7.3, we can now compute the expected accuracies of each of the five paths from RS to FT. If the individual operation accuracies are independent of each other, then the path accuracy is simply each of the edge accuracies in the path multiplied together. These are shown in Table 7.2

Table 7.2   Pathways through the graph of Figure 7.3, with the edge probabilities and the total probability along the path (all of the edge probabilities multiplied; i.e., the assumption of independence). The final column is the rank by accuracy of the path (highest probability). In this case, the longest path, RS-FS-ES-ET-FT, has the highest predicted accuracy of 0.9029.

Path	Edge probabilities	Total probability	Rank
RS-RT-FT	(0.95)(0.94)	0.8930	2
RS-RT-ET-FT	(0.95)(0.93)(0.99)	0.8747	3
RS-ES-ET-FT	(0.88)(0.99)(0.99)	0.8625	5
RS-FS-FT	(0.94)(0.93)	0.8742	4
RS-FS-ES-ET-FT	(0.94)(0.98)(0.99)(0.99)	0.9029	1

In Table 7.2, we find that the longest path – that is, the path with the most operations – has the highest expected accuracy, even though it has at least one more step than any other path. This is because each operation in this path, RS-FS-ES-ET-FT, has high (0.94) or very high (0.98–0.99) accuracy. The overall expected accuracy is 0.9029, compared to 0.8930 for the much simpler path RS-RT-FT.

There are a number of concerns, or at least caveats, about the example just presented. The first is that the different operations are not likely to be independent of each other. For example, if a sample of Russian speech is difficult to translate into French speech, it is likely that it is also difficult to translate into English speech, and maybe even to transform into Russian text. Thus, several of the probabilities listed on the edges may be correlated. The second concern with the example is that there is no penalty for the number of operations performed: RS-FS-ES-ET-FT has four steps, while RS-RT-FT has only two steps. This could mean, for example, that RS-FS-ES-ET-FT costs twice as much to perform as RS-RT-FT, takes twice as long to perform (since it requires twice the operations), and/or is much more sensitive to changes in the inputs (and is thus less robust to data drifting).

On the other hand, having the analysis path pass through either English speech (ES) or English text (ET) nodes may have some particular advantages. Since English, based on the accuracies reported, is the "central" language for the overall system of text analytics operations, having the input content internalized as English may be highly advantageous for the repurposing of the content. Suppose that another language (e.g., Spanish or Mandarin Chinese) or another application (e.g., summarization or document clustering) is added to the system. Having the ES/ET information also allows the data analyst concerned with testing and configuration to have a *lingua franca*, as it were, for comparing two different systems. If every major text analytics task to be performed is channeled through ET/ES, then the ET/ES data sets can be "fairly" compared to one another for selecting an optimum system configuration. That is, the ET/ES "central" content is what can be used for benchmarking one system configuration versus another.

Irrespective of its overall advantages, the method shown in this section can be used to compare and contrast different pathways for multi-step text analytics tasks. Here, the function being evaluated is the optimum pathway for an important systems metric such as accuracy.

7.2.2 Text Analytics and Immunological Data

In the previous section, we discussed the concept of a central language in a multi-lingual text analytics system. Here, we make the case for functional text analytics itself to be the "central language" for a wide variety of analytics approaches. This is different from the use of vehicle counting as a calibration proxy for tracking accuracy described above: that example is simply the application of a functional approach to an analytics field other than text. Instead, the use of text analytics as the central analytics approach means that specific forms of text analytics are applied to other fields.

In this section, text analytics approaches are repurposed to be used as functional approaches for the analysis of immunological data. Summarization, clustering, and classification will be discussed briefly here for their applicability to the analysis of immunological data. For summarization, the Fab (variable) fragment amino acid sequences in antibodies can be assessed using the same "total sentence score" technique as given in Chapter 2 (Equation (2.12)). The key is to determine what the words are. For the variable fragments of antibodies, amino acid heavy and light chain residues are typically on the order of 200 amino acids in length [Rodr98]. There are 20 amino acids, and so this means these residues are effectively 200-character length words with a 20-character alphabet. These heavy and light chains are composed of variable and constant domains with lengths as short as 6 amino acids, making the alphabet and word lengths of these residues relatively similar to spoken languages. The residues comprising a population of sequenced Fab fragments can therefore be assigned weights, and the summarizations of these sequences be computed from the set of "words" comprising each Fab fragment.

Clustering of immunological data can proceed using the amino acid sequences of the constant and variable portions of the residues. The application of regularization to the clustering can also be applied, although a subtle difference from that of Chapter 3 is appropriate here. Here, the basis for clustering is a regularized combination of the constant sections and the variable sections, as shown in Equation (7.1).

$$\text{Clustering Decision} - \lambda(\text{Constant Clustering})$$
$$+ (1 - \lambda)(\text{Variable Clustering}). \qquad (7.1)$$

The first term, λ(Constant Clustering), indicates that λ proportion of the cluster assignment, where $0 < \lambda < 1$, is due to the constant portions of the antibody residues, while $(1 - \lambda)$ proportion of the cluster assignment is due to the variable portions of the antibody residues. When λ is close to 1.0, the clustering is essentially aligned with the constant portions so that the normal biological classifications of antibodies, such as IgA, IgE, and IgG, will be observed. However, when λ is close to 0.0, the clustering is largely due to the variable portions, and so antibodies will be clustered based on what antigens they will bind to rather than what type they are. Thus, the regularization is used to control the balance between antibody class specificity and antigen type specificity.

For classification, the approach outlined for clustering suffices to distinguish between immunoglobulin (Ig) classes, so long as λ is appreciably greater than 0.0. However, in most cases, a class is defined by a training set. In order to assign the antibody sequences to a class, bioinformatics are used [Grom10]. However, the text classification approach can be used for immunological classification as well. Homologous sequences can be used in a manner like synonyms, although this makes more sense for DNA or RNA than for proteins (since on the mean, there are more than three codons for each amino acid; thus, a mean of three "synonyms" for every amino acid). This is an important point since the approaches for summarization, clustering, and classification applied to antibodies can be applied not only to other protein sequences but also to DNA and RNA sequences of any form.

7.2.3 Text Analytics and Cybersecurity

Text analytics are also useful for cybersecurity analytics. Text analytics approach like TF*IDF is readily applied to network packets in order to cluster similar packets (e.g., similar headers, payloads, and/or trailers) and, in particular, suspicious packets for the analysis of their origin, delivery route, and/or target. Network traffic can also be analyzed using text analytics approaches. For example, a distributed denial of service (DDoS) attack will show different origin distribution (lots of traffic from lots of places, all targeting the same destination) than a port scan (lots of traffic from one place, corresponding to a host discovery attack). If locations are the words, and the packets from the locations word occurrences, different types of attacks can be categorized with the text analytics of histogram analysis and TF*IDF. Outgoing network traffic can also be analyzed with an analogous approach.

7.2.4 Data-Ops and Testing

Data operations are the specific text analytics to be performed, along with their sequence. It is the plan for data collection, aggregation, and analysis during the operation of the system. Therefore, it is also the plan for data collection, aggregation, and analysis during testing and definition of system configuration. As Jimmy Carter noted, "Testing oneself is best when done alone," and for our purposes, this means that the personnel performing the testing should be able to proceed without further input from the development team. Like Eartha Kitt ("I have to keep testing myself"), the testing team has ownership for exploring the limitations and vulnerabilities of the project, and their approach should not be disrupted by the research or development teams. The Data-Ops are agreed on as the project starts, and testing is done in parallel with development.

7.3 Machine Learning Aspects

We can use text analytics to complement image, video, and other non-text machine learning. Text can be directly associated with an image through metadata such as file information, author information, and other indexing information. Text statistics can also be generated in the form of descriptors for the individual images or frames. Also, the operations performed on the images are represented as text. Metadata about non-text information such as time-series data and images can also include any operations performed on the data; for example, enhancement, restoration, or compression.

For testing and configuration, machine learning is most important as a component in the overall analysis of how the system performs. Machine learning algorithms continue to change as more input is created for the associated training, validation, and other testing checkpoints. This creates a chicken-and-egg scenario in which testing must be performed on as much of the machine learning related training data as possible, while simultaneously directing the need for specific new training data based on the findings of the test team. This might appear to contradict the need for independence of the testing team as recommended in the previous section; however, it is to be remembered that the testing team is the team responsible for soliciting new training content, and, in this way, their independence of action is ensured. It is the test team that is closing the loop on the creation of test content; the development team is not determining the need or process of creation for new content.

From this perspective, a machine learning element in a text analytic system is a functional element. Testing and the determination of the system's optimum configuration impact the settings for the machine learning elements, even as the machine learning elements affect the output of the testing. This is logically cohesive with the explanation of the general form of functional text analytics described earlier in this chapter, wherein using a second process to determine the settings for the first process is shown to create a more robust system.

7.4 Design/System Considerations

Testing and system configuration are core elements of a system's design. Testing is the means, in fact, of determining what the settings in the system configuration should be. Humans are linguistic, and, thus, text analytics are central to understanding content in any communication media. In several of the linguistic system discussed in this book, English was determined to be the "central" language of a multi-language system. This was determined heuristically, when it was noticed that the use of English as an intermediate language led to improved overall system accuracy, even when one or two extra operations are required.

In addition to the heuristic approach, there is a non-heuristic means of assessing the central language in a multi-language system which is described here. In Table 7.3, the matrix of error rates between languages in an operation is given. This matrix is for text-to-text translation for the EFIGS (English, French, Italian, German, and Spanish) languages. The translation accuracy from English to French, for example, is 0.99. The opposite direction, that of translating from French to English, has a similar but lower accuracy of 0.97.

The rows and columns of Table 7.3 provide some valuable insight into the working of the overall multi-language system. Taking the mean of the rows, for example, we see that when English is the source text, the mean translation accuracy is 0.985. For the other languages, this is substantially lower: 0.96 from French, 0.96 from Italian, 0.943 from German, and 0.958 from Italian. These clearly group with English being easily the most accurate, German the least accurate, and the three Latin languages intermediate in accuracy. Taking the mean of the columns, translations into English have a mean

Table 7.3   Translation accuracies from the source language (first column) to the destination languages (columns 2−6). Please see text for details on analyzing centrality of the system.

Source	To English	To French	To Italian	To German	To Spanish
From English	N/A	0.99	0.98	0.99	0.98
From French	0.97	N/A	0.96	0.95	0.96
From Italian	0.95	0.96	N/A	0.96	0.97
From German	0.96	0.94	0.95	N/A	0.92
From Spanish	0.95	0.97	0.98	0.93	N/A

Table 7.4 Translation accuracy ratios to/from the other languages (ratios of To and From data in Table 7.3), computed to determine if translation asymmetries exist. Please see text for details on analyzing the asymmetry of the system.

Source	English To/From	French To/From	Italian To/From	German To/From	Spanish To/From
English	N/A	1.021	1.032	1.031	1.032
French	0.980	N/A	1.000	1.011	0.990
Italian	0.969	1.000	N/A	1.011	0.990
German	0.970	0.989	0.990	N/A	0.989
Spanish	0.969	1.010	1.010	1.011	N/A

accuracy of 0.958. The means for translating into French, Italian, German, and Spanish are 0.952, 0.954, 0.946, and 0.946, respectively. While English is again the highest mean accuracy for the columns, the differences between the columns are much less (less than a third) than for the rows. Thus, the differentiating accuracy for the entire system is the accuracy of English text into the other four languages. The lower accuracy of translating German text into the other four languages is also a characteristic of the system. Combined, these results indicate that English is the **central language** for the system and that German is probably the language for which the system builders had the least expertise. However, since French, Italian, and Spanish are more closely related in syntax and vocabulary, it is possible that this similarity collectively lifts their results above those of German.

In order to address whether or not the system has asymmetry, we take the ratios of "To/From" for each language pair. For English and French, then, the "To/From" ratio for English is 0.99/0.97 = 1.021. For French, it is the inverse, 0.97/0.99 = 0.980. These ratios are collected in Table 7.4. As in Table 7.3, this table contains 20 relevant values (the diagonal is "not applicable" or N/A).

The rows of Table 7.4 are analyzed using a simple z-value, computed in Equation (7.2).

$$z = \frac{|\mu - 1.0|}{\sigma/\sqrt{n}}. \tag{7.2}$$

Here, $n = 4$ since there are four values for each row. The mean of the row, μ, is the mean of the four non-diagonal values, and the standard deviation of these fours values is σ. The p-values (two-tailed since we do not know *a priori* whether the values are to be greater or less than 1.0) of the z-scores are shown, along with μ, σ, and the z-score in Table 7.5.

The results of Table 7.5 show that both English and German have asymmetrical behavior. English as a language translated to another language has higher accuracy than English as a language translated from another language. German has the opposite behavior. This asymmetric behavior is a

Table 7.5   Calculation of asymmetry. The *z*-value is computed according to Equation (7.2), and the *p*-value is calculated from a *z*-table (two-tailed test). If $p < 0.05$, then the language is considered asymmetric. In this table, English is positively asymmetric while German is negatively asymmetric. Please see text for details.

Language	Mean To/From	Std. To/From	*z*-value	*p* (*z*-value)
English	1.029	0.005	10.83	0.000
French	0.995	0.013	−0.71	0.475
Italian	0.993	0.018	−0.84	0.401
German	0.985	0.010	−3.20	0.00136
Spanish	1	0.021	0	1.000

form of sensitivity analysis for the linguistic system. Any such asymmetries are indicative of overall system immaturity, meaning that there is room for improvement in the overall system accuracy. Thus, the possibility of linguistic asymmetries should always be investigated. In the current system, however, it means that given the choice for a pipeline, we would prefer to move from English text and to German text as steps in a pipeline. This is because these steps have asymmetrically higher accuracy than their opposites, moving to English text and from German text.

The approach outlined in this short section is concerned with translation, but it could also be used for any other multi-stage text analytics process, including one extending from key words to summaries to documents to clusters of documents. The central analytic will be the one with the highest accuracy, and asymmetries in the steps between two types of data allow us to determine preferential elements in our processing pathways.

7.5 Applications/Examples

Functional text analytics extend to multimedia. The first and perhaps the most obvious one is video. Multi-lingual video requires subtitles, the relative value of which can be determined with minimal human intervention. For subtitles, tests of comprehension can be given to watchers of the original-language video for comparison to watchers of the subtitled videos. These can be made very simply, using Likert, multiple choice, and True/False questions. The use of closed captioning can also be readily evaluated by comparing the comprehension of those using closed captioning with those listening only (caveat there are perhaps two differences in input here, the presence/absence of text and the potential translation). This approach can readily be extended to virtual/augmented reality (AR/VR), where the text is associated with the environment in which the person is interacting.

Table 7.6 Example of a feature A that provides a good proxy for feature X. At each of the settings for X, the behavior for A is consistent (in this case, positively correlated). The converse is true since at each of the settings for A, the behavior for X is consistent (in this case, positively correlated).

Accuracies	Setting A1	Setting A2	Setting A3	Setting A4
Setting X1	0.76	0.79	0.84	0.88
Setting X2	0.79	0.84	0.88	0.89
Setting X3	0.81	0.85	0.91	0.93
Setting X4	0.82	0.86	0.91	0.95

Table 7.7 Example of a process A that does not provide a good proxy for process X. No consistent relationship between processes A and X occurs for different settings of the two processes.

Accuracies	Setting A1	Setting A2	Setting A3	Setting A4
Setting X1	0.76	0.84	0.81	0.78
Setting X2	0.79	0.83	0.85	0.88
Setting X3	0.84	0.79	0.78	0.83
Setting X4	0.76	0.82	0.79	0.76

For testing and configuration determination, the system designer is particularly interested in sensitivity analysis of the different variables at her avail for the system design. One of the advantages to functional text analytics approaches are that they, in general, equate sensitivity in design settings with the sensitivity in the features used for the specific proxy metrics. As argued earlier in the chapter, system robustness is one of the primary reasons for using a proxy process to tune a specifically desired process. From a statistical design standpoint, trading off one proxy output for another is only recommended when the behavior of the proxy system is "interchangeable" with the behavior of the particular system of interest. This is illustrated by the data in Tables 7.6 and 7.7. In Table 7.6, four different sequential settings for the process of interest, A, are used in conjunction with four different sequential settings for putative proxy process, X. The settings are highly correlated in Table 7.6, with both rows and columns varying in the same manner irrespective of the settings. Here, process A is deemed a good proxy for process X for this particular data set.

In Table 7.7, however, the rows and columns do behave the same across the different settings. As such, process A is not deemed a good proxy for process X in this data set.

Other equally straightforward calculations can be performed that are of value for testing different candidate configurations. For example, the variance or the ratio of variance to mean (coefficient of variance or COV) is a good test for system stability. This is well-known to engineers and technicians working in the quality assurance (QA) field – a batch of products with the same mean but a higher variance is almost always lower quality than the batch with the same mean and lower variance. The higher variability corresponds to lower reliability in any circumstance in which the mean value is above the specified minimum value for quality since, by the definition of variance, a larger part of the distribution will then be below the minimum quality value. For example, suppose the minimum quality is 100 in some measurement. Two batches, each with mean qualities of 120 but standard deviations (square root of the variance) of 20 and 10 are observed. Assuming Gaussian distributions, the batch with a standard deviation of 20 is expected to have 16% of its samples below 100, while the batch with a standard deviation of 10 is only expected to have 2.5% of its samples below 100. Clearly, one out of six products failing QA is a lot less reliable than 1 out of 40 failing QA.

This definition of reliability readily extends to text analytics. Writing style, including the assessment of authorship, potential plagiarism, or ghost writing, can be assessed through a form of QA on the content in multiple text elements. One means of assessing the relative quality of a writer is by removing the proper nouns and then tabulating frequency histograms of salient words in the language. Salient words are words that are not overly common (such as "the," "or," and "it") but also not overly rare (such as "chromosome" or "zymurgy"). Intermediate frequency words (such as "magazine" or "firewood") might be words that are between the 1000th and 10,000th most frequently occurring words in a language. For this set of words, the mean and standard deviation (std.) of occurrence frequencies (percentage of overall words in a document) are computed and then the z-value from a given document to each of these means is computing using the z-score of Equation (7.3):

$$z = \frac{|\mu - \mu_{\text{document}}|}{\sigma/\sqrt{n}}. \tag{7.3}$$

Here, the mean μ is the mean frequency of occurrence of the term in all documents, and the standard deviation σ is the standard deviation of occurrence of the term in all documents. For a given document, μ_{document} is the frequency of occurrence of the term in the document, and $n = 1$ since it is one document. Three example distributions are shown here, and note that the standard deviation is generally relatively high since the terms are relatively infrequent:

Word A: mean frequency $\mu = 1.43 \times 10^{-5}$, std. $\sigma = 1.13 \times 10^{-5}$
Word B: mean frequency $\mu = 7.43 \times 10^{-6}$, std. $\sigma = 5.44 \times 10^{-6}$
Word C: mean frequency $\mu = 5.39 \times 10^{-6}$, std. $\sigma = 6.11 \times 10^{-6}$

It should be noted that this is a parametric approach to authorship determination; that is, one based on the actual word frequency data itself. The sum of z-scores for these words in a document will tend to be the lowest when the document was written by the same author as the training set. Assuming a large enough training set, the lower the difference, the more **reliability** in assigning the authorship.

An analogous approach can be pursued with non-parametric data as well. In the non-parametric case, the 1000th to 10,000th most common words in the document's language are ranked in order of their frequency in the document and compared to their order in other documents by different potential authors. The ranked order of terms from authors most similar to those in the document will tend to be most similar to the ranked order in the document. Differences in ranked orders are illustrated elsewhere in this book; for example, see weighted differences in reading orders (Equation (6.2)) in the previous chapter. For our purposes here, again assuming a large enough training set, the lower the difference in this non-parametric comparison (just as for the parametric case), the more **reliability** in assigning the authorship.

7.6 Test and Configuration

Speaking of assessing authorship, perhaps you, the reader, are wondering if this section has been written by the Department of Redundancy Department since this is the "test and configuration" section of a chapter with the same name. However, there has been a "test and configuration" section in every chapter of this book, so we are not going to stop here. Instead, we talk about how a test and configuration decision for a text analytics application can itself be tested and configured. There are three relatively simple thought experiments, or *Gedankenexperiments*, we introduce here as means of such an assessment.

(1) How does your test and configuration interact with temporal delay (the "Mars mission" *Gedankenexperiment*)?
(2) How will your system help us know what the culture surrounding the text was like to someone 100 years from now (the "black hole" *Gedankenexperiment*)?
(3) Can this functional text analytic approach be extended to other, non-text, domains (the "interstellar" *Gedankenexperiment*)?

Each of these three *Gedankenexperiments* will now be briefly considered.

The first *Gedankenexperiment* is the "Mars mission." This scenario involves assuming that there is a delay between when your system is put in place and when meaningful data can be collected. When a spaceship reaches Mars, somewhere between 55 and 400 million kilometers away from Earth, it takes light – and more importantly for our purposes, radio waves – between 6 and 45 minutes to travel from Earth to Mars and back. This means that you have to build your mission to accommodate up to a 45-minute communication delay. In the same way, for a text analytic system, we have to build in a delay between when training occurs and when the training data is updated. A robust, systems-engineering based design will use the feedback of successfully completed operations to replenish the training data. This has been designated the "Proof by Task Completion" feedback pattern [Sims13] and is an important part of keeping the system settings current. When a system is being designed, it is very important to outline how much of a delay between settings and feedback on settings can be tolerated. After all, the best time to send people to Mars is when it is going to be 55 million kilometers away. The best time to update the data used to define system settings is whenever this update will result in a measurable, positive effect on the system's behavior (cost, performance, accuracy, robustness, resilience, ability to be performed in parallel, etc.). This ties the Mars mission to specific functionality.

The second *Gedankenexperiment* is the "black hole." A familiar – or should we say clichéd? – plot for a sci-fi story is to have black holes be useful as portals in space-time, with travelers into the black holes emerging at a very different time in the continuum. Thinking about how information might have to travel through time, we must be assiduous about including contextual information sufficient to be able to recreate the "cultural environment" in which text information was created. One of the frustrations of parenting or teaching is not as much the gulf between what facts the educator has that the student does not, but all of the subtleties around the facts. These days, often for admirable reasons, it is fashionable to judge people in the past based on the standards of today. Similarly, we will be judged in the future for atrocities we are allowing to happen. Being able to provide as much contextual information, or "metadata," as is required to recreate the "feel" for the text analytics may be just as important as the context itself. Very old literature – think of *Gilgamesh*, the *Dead Sea Scrolls*, and the *Book of the Dead* – suffers from lack of context that would be invaluable to scholars today. For your black hole thought experiment, imagine that you wish to convey to your readers a hundred years from now what the text meant to you in context of everything else that is happening outside of the text. After all, context means "with text," and a proper text life cycle design will include as much context as necessary to allow the reader to recreate the environment in which the text lived. A black hole thought experiment brings the functionality of greater contextual understanding.

The third *Gedankenexperiment* is the "interstellar." A few years back, Oumuamua passed through our solar system [Tril18] and was the first confirmed interstellar object making such a journey. Functionally, then, Oumuamua was repurposed from its role in some other solar system (maybe a chunk of a planet's crust, for example) to its role as a potential meteorite in our solar system. The interstellar thought experiment is one in which you imagine how the text that you are producing might be reused later. Such reuse is termed *repurposing*. Music sampling is an example of repurposing. Text sampling, however, might be called plagiarism, but, fortunately, it is not what we are most concerned with for functional text analytics. Instead, we are concerned with functional representations of text that can be used for purposes outside of the purpose of the original text. A translation, for example, is a functional representation of a text element. Different languages have different figures of speech, idiomatic expressions, and specialized terminologies (including slang and jargon). In order to make text repurposable under a variety of unanticipated circumstances, a different form of text metadata that provides disambiguation for the text is recommended. For example, consider the famous line "To be or not to be, that is the question." In order to make this line repurposable at any point in the future, an example metadata representation of the line might go as follows: "To be [to continue living] or not to be [to discontinue living, through one's own volition, as in suicide], that [the choice to live or take one's own life] is the question [as in "the" question, the central question, the most important question to decide]." This may seem obvious, but an interpretation of the line as "To be [decisive, as in killing Claudius] or not to be [as in vacillating as Hamlet has been doing up to this point in the play], that is the question" is not unreasonable in the context. The point is that the richer the metadata, the more useful the content. Metadata begets functionality, and we are avid fans of that.

Our last chapter-opening quote in this book is one familiar to readers of our other books on analytics and algorithms [Sims13][Sims19]. Text analytics are useful for prediction, in addition to the more functional applications highlighted in this book. But all predictive analytics have a (hopefully tolerable) degree of uncertainty associated with them, and so the future is indeed uncertain. The only thing certain in this world is uncertainty, and analytics are a means of managing the uncertainty through reducing its probability and its severity. To that list, we can add one more certainty, however. The end (of this book, at least) is indeed near. Thanks for reading!

7.7 Summary

This chapter describes testing and measurement aspects of functional text analytics, and the corresponding impacts of these on how text analytics systems are configured. The process by which ground truthing in one application is replaced with simpler, most cost-effective, or even more ethical ground truthing in another field was described. Its extension to other fields was illustrated with an example involving video analytics. The principle of finding the proxy training approach with the highest correlation to the desired system output was presented. The use of generalized graphs to determine the optimum pathway of operations to perform for a given text analytics task was illustrated through an example in which Russian speech was translated into French text. Using English as a central language for linguistic analytics was revisited here. Then, using text analytics as a central means for other analytics was applied to immunological informatics with summarization, clustering, and classification highlighted. Application of text analytics to cybersecurity, and the need for an independent test team, was highlighted. A brief consideration of multi-lingual systems followed, in which the means to compute the central language along with any asymmetries in translation behavior was overviewed. The chapter concluded with some pragmatic means of assessing good proxy processes, system reliability, several *Gedankenexperiments*, and the extension of functional text analytics to other fields.

References

[Grom10] Gromiha MM, "Protein Bioinformatics: From Sequence to Function," Academic Press, 2010.

[Rodr98] Rodríguez-Romero A, Almog O, Tordova M, Randhawa Z, Gilliland GL, "Primary and Tertiary Structures of the Fab Fragment of a Monoclonal Anti-E-selection & A9 Antibody That Inhibits Neutrophil Attachment to Endothelial Cells," J Biol Chem **273** (19), pp. 11770-11775, 1998.

[Sims13] Simske S, "Meta-Algorithmics: Patterns for Robust, Low-Cost, High-Quality Systems," Singapore, IEEE Press and Wiley, 2013.

[Sims19] Simske S, "Meta-Analytics: Consensus Approaches and System Patterns for Data Analysis," Elsevier, Morgan Kaufmann, Burlington, MA, 2019.

[Spär72] Spärck Jones K, "A Statistical Interpretation of Term Specificity and Its Application in Retrieval," Journal of Documentation **28**, pp. 11–21, 1972.

[Tril18]   Trilling DE, Mommert M, Hora JL, Farnocchia D, Chodas P, Giogini J, Smith HA, Carey S, Lisse CM, Werner M, McNeill A, Chesley SR, Emery JP, Fazio G, Fernandez YR, Harris A, Marengo M, Mueller M, Roegge A, Smith N, Weaver HA, Meech K, Micheli M, "Spitzer Observations of Interstellar Object 1I/'Oumuamua," The Astronomical Journal 156:261 (9 pp.), 2018.

Index

About the Author

Steve Simske
Steve Simske is a Professor with the Systems Engineering Department, Colorado State University (CSU), Fort Collins, CO, USA, where he leads research on analytics, cybersecurity, sensing, imaging, and robotics. He was a former Fellow in HP Labs. He was in the computer and printing industries for 23 years before joining CSU in 2018. He holds more than 200 US Patents and has more than 450 publications. This is his fourth book.

Marie Vans
Marie Vans is a Senior Research Scientist with HP Labs, Fort Collins, CO, USA. She is currently working with the AI & Emerging Computer Lab, where she is focused on developing virtual reality simulations for education, product introduction, and analytics associated with educational experiences in virtual reality. She is also on the faculty of the San José State University, School of Information. She holds a Ph.D. and M.S. in computer science from Colorado State University, Fort Collins, CO, USA, an MLIS from San José State University, San Jose, CA, USA, and has more than 55 published papers and 35 US granted patents.